Lactobacillus - A Multifunctional Genus

Edited by Marta Laranjo

Published in London, United Kingdom

Lactobacillus - A Multifunctional Genus
http://dx.doi.org/10.5772/intechopen.98170
Edited by Marta Laranjo

Contributors
François Coutte, Benoit Cudennec, Cyril Raveschot, Morgan Le Rouzic, Pauline Bruniaux, Vincent Phalip, Rozenn Ravallec, François Krier, Rodney H. Honrada Perez, Amily E. Ancuelo, Parul Thapar, Mohinder Kumar Salooja, Pedro Maldonado-Alvarado, Ederson Freire-Almeida, Nagina Rafiq, Saiqa Bashir, Imtiaz Hussain, Imran Hayat, Turfa Mamoona, Christopher Osita Eze, Dinebari Phillip Berebon, Thaddeus Harrison Gugu, Francis Ifeanyi Anazodo, James Ekemezie Okorie, Arekal Nagaraja Roopashri, Janakiraman Savitha, M.S. Divyashree, B.S. Mamatha, K.A. Usha Rani, Ashwini Kumar

First published in London, United Kingdom, 2023 by IntechOpen
IntechOpen is the global imprint of INTECHOPEN LIMITED, registered in England and Wales, registration number: 11086078, 5 Princes Gate Court, London, SW7 2QJ, United Kingdom

British Library Cataloguing-in-Publication Data
A catalogue record for this book is available from the British Library

Additional hard and PDF copies can be obtained from orders@intechopen.com

Lactobacillus - A Multifunctional Genus
Edited by Marta Laranjo
p. cm.
Print ISBN 978-1-80355-444-0
Online ISBN 978-1-80355-445-7
eBook (PDF) ISBN 978-1-80355-446-4

Meet the editor

Marta Laranjo holds a Ph.D. in Biology and is a senior researcher at MED - Mediterranean Institute for Agriculture, Environment and Development (Food Science and Technology Group), Universidade de Évora, Portugal. She is also the executive director of the UNIMED Subnetwork on Food and Water. An expert on microbiology and molecular biology, her main interests include food microbiology, with a special emphasis on starter cultures and the dynamics of food microbiota. She has published several book chapters and numerous research papers in international peer-reviewed journals. She is also a frequent reviewer for several journals in the areas of microbiology and molecular biology. Dr. Laranjo has supervised several Ph.D., Master, and undergraduate students.

Contents

Preface

This Edited Volume is a collection of reviewed and relevant research chapters, concerning the developments within the *Lactobacillus* field. The book includes scholarly contributions by various authors and edited by a group of experts pertinent to *Lactobacillus*. Each contribution comes as a separate chapter complete in itself but directly related to the book's topics and objectives.

The book includes chapters dealing with the topics:

Chapter 1 "Indian Traditional Fermented Foods: The Role of Lactic Acid Bacteria"

Chapter 2 "*Lactobacillus* Use for Plant Fermentation: New Ways for Plant-Based Product Valorization"

Chapter 3 "Diverse Bioactive Molecules from the Genus *Lactobacillus*"

Chapter 4 "Bacteriocins: Applications in Food Preservation and Therapeutics"

Chapter 5 "Use of *Lactobacillus* for Lactic Acid Production from Agro-Industrial By-Products"

Chapter 6 "Lactobacilli: Application in Food Industry"

Chapter 7 "*Lactobacillus* exopolysaccharide: An Untapped Biopolymer"

Chapter 8 "Effect of Sodium Acetate and Trace Element (Se^{2+}, Zn^{2+}) on Exopolysaccharide Production by *Lactobacillus plantarum* and Promote Antioxidant Capacity"

IntechOpen

Chapter 1

Indian Traditional Fermented Foods: The Role of Lactic Acid Bacteria

Arekal Nagaraja Roopashri, Janakiraman Savitha, M.S. Divyashree, B.S. Mamatha, K.A. Usha Rani and Ashwini Kumar

Abstract

Fermentation technology is an important field comprising the use of microorganisms and enzymes to produce the compounds that have applications in the food, pharmaceutical, energy, and chemical industries. Although food fermentation processes have been used for generations as a prerequisite for sustainable food production, today it has become more demanding to obtain functional and therapeutic food products through the application of continuous creations and advancement of innovative fermentation processes. For these reasons, efforts are directed toward designing new processes, concepts, and technologies. Fermentation is a natural process whereby microorganisms such as lactic acid bacteria and yeast convert carbohydrates such as starch and sugar into alcohol or acids, both of these act as a natural preservatives. This process is still used today to produce foods such as wine, cheese, sauerkraut, yogurt, and other types of traditional foods. Traditional fermented foods are popularly consumed and form an integral part of our diet since early history. They are recognized as having multiple benefits related to nutritive values, therapeutic properties, and sensory attributes. In most fermented foods, the fermentation process is predominantly initiated by lactic acid bacteria. These organisms have been termed as probiotic bacteria—a group that appears to have specific health-promoting attributes.

Keywords: lactic acid bacteria, probiotics, prebiotics, traditional fermented foods, human health

1. Introduction

The production of fermented foods is one of the oldest food processing technologies well known to mankind. Since the beginning of civilization, methods of fermenting milk, cereals, legumes, vegetables, and meats have been described [1]. The preparation of these kinds of fermented foods will be with us in the far future, as they are a source of alcoholic foods/beverages, vinegar, pickled vegetables, sausages, cheeses, yogurts, vegetable protein amino acid/peptide sauces, and pastes with meat-like flavors, and leavened and sourdough breads, and so on. Fermented foods are of great importance because they provide and preserve an enormous amount of nutritious foods with a wide variety of flavors, aromas, and textures that enrich the human diet. As used herein, the term "nutrition" or "nutrient" will include providing the consumer with calories/energy, protein, essential amino acids/peptides, essential fatty acids, vitamins, and mineral requirements that contribute to the solution of nutritional problems and diseases in the human population [2].

In most cases, the procedures and knowledge associated with the manufacture of these food products were passed on from generation to generation within local communities, monasteries, and medieval farms. In the mid-nineteenth century, two events occurred that had a significant impact on our understanding of the method and process of food fermentation. First, a huge number of residents from the villages started moving toward the towns and cities due to the more opportunities in industrial sectors. Hence, the practicing of conventional methods to prepare food for more population was no longer operative. This led to inventing new processes for the manufacturing of vast quantities of food products, which demanded the industrialization of the food sector. Second, the progress in the field of microbiology in the 1850s led to an understanding of the basic science of the fermentation process for the first time. Consequently, the important function of lactic acid bacteria, yeasts, and molds in the production of fermented foods was understood, which eventually led to a more controlled and efficient fermentation process [3].

For many fermented foods particularly dairy products, the characterization of microorganisms that are responsible for fermentation is very important for their usage in the dairy industry. Therefore, in the late nineteenth-century isolation of starter cultures and manufacturing on a large scale was initiated to supply to the factories involved in the manufacture of dairy products [3]. Generally, starter culture strains are often used to improve the nutritive value and sensory characteristics of fermented foods, maintain safety and quality, and promote nutrition [4]. In recent years, it has been a great strategy in food microbiology to study microorganisms with various functions, to use as a starter culture [5]. Several studies have reported the preparation of food products using multifunctional microorganisms, wherein lactic acid bacteria (LAB) are particularly recommended as starter cultures for the fermentation process due to their benefits in terms of probiotic properties, antimicrobial production, beneficial enzyme production, and enhancement of other functionality [6–11]. They are normally considered safe and widely used as a starter culture in the production of fermented foods [12]. Due to the unique metabolic characteristics, they are involved in many fermentation processes of milk, cereals, vegetables, and meats. They are effective as probiotics and exhibit beneficial properties such as the production of antimicrobial compounds, enzymes, involvement in immune regulation, and antioxidant activity [13]. The microbiota in the human gut displays significant influence on host immunity, nutrition supply to the body as well as a contribution toward physiological function, whereas the enhanced therapeutic role of beneficial microbes in the gut is mainly due to the improvement of their population in intestinal microbial communities and their subsequent correlation with human physiology and disease pathogenesis [14]. In 2006, FAO/WHO defined probiotics as "live microorganisms, which upon ingestion in adequate amounts confer a health benefit to the host [15]." Hence, probiotics are considered as safe products for the host when consumed. The key criteria for the selection of probiotics which are mainly based on the FAO/WHO guidelines include safety (nonpathogenic strains without toxicity), resistance to gastric and bile acids, adhesion to epithelial cells, and antimicrobial activity (antagonism against pathogens). These guidelines recommend performing *in vivo* experiments on strains that have demonstrated potential health benefits based on *in vitro* experiments. In addition, probiotic properties are strain-specific, and hence, each strain characteristic needs to be verified [16].

DOI: http://dx.doi.org/10.5772/intechopen.110741

The lactic acid bacteria (LAB) are considered as probiotics and play important role in the preservation and production of healthy fermented food products. The genera of LAB consist of *Lactobacillus*, *Leuconostoc*, *Lactococcus*, *Enterococcus*, *Pediococcus*, *Streptococcus*, *Weisiella*, etc. The genus *Lactobacillus* includes 261 species that differ greatly in their phenotypic, genotypic, and ecological properties. Recently, [17] proposed reclassification of lactic acid bacteria into 25 genera. The phylogeny is based on whole-genome sequences, which includes the revised genera *Lactobacillus* and *Paralactobacillus*. The remaining 23 genera are novel: *Holzapfelia*, *Amylolactobacillus*, *Bombilactobacillus*, *Companilactobacillus*, *Lapidilactobacillus*, *Agrilactobacillus*, *Schleiferilactobacillus*, *Loigolactobacilus*, *Lacticaseibacillus*, *Latilactobacillus*, *Dellaglioa*, *Liquorilactobacillus*, *Ligilactobacillus*, *Lactiplantibacillus*, *Furfurilactobacillus*, *Paucilactobacillus*, *Limosilactobacillus*, *Fructilactobacillus*, *Acetilactobacillus*, *Apilactobacillus*, *Levilactobacillus*, *Secundilactobacillus*, and *Lentilactobacillus*. Those LAB belong to the homo- and hetero-fermentative groups and generally require enriched artificial media. They grow naturally on most food substrates which led to the lowering of pH to a point where other competing microorganisms are unable to grow. For example, *Lactococcus* and *Leuconostoc* normally lower the pH to 4.0–4.5, whereas *Lactobacilli* and *Pediococci* lower the pH to 3.5 as reported by [18]. Several research studies have shown that the use of LAB as fermentation agent for the preparation of food products is considered as safe for consumption [7, 10, 11].

It is well-known fact that India is recognized for its rich traditional fermented food products. Generally, in the Indian subcontinent, fermented foods are prepared using indigenous food crops and other biological resources. Hence, the nature of food products and their base materials varies in each province. Presently, there are a number of fermented food products with different base materials and preparation methodologies have been reported. Each fermented food is allied with a unique group of microbiota, which has the potential to increase the level of proteins, vitamins, essential amino acids, and fatty acids. In local communities, the preparation of traditional fermented foods is still done by spontaneous fermentation method and their microbiota profile get varies each time. Hence, limited knowledge is available about the microbiota of these kinds of food products [19, 20].

This chapter outlines the role of lactic acid bacteria in food fermentation, their probiotic properties (gastrointestinal tract acidic tolerance, adhesion, hydrophobicity, and auto-aggregation), multifunctional characteristics (antimicrobial, antioxidant, phytase, and β-galactosidase activities), and the mechanisms of antagonistic nature of LAB as well as a brief explanation of some significant traditional fermented foods from India. It is expected that more research needs to be carried out toward strain improvement with beneficial properties in order to use in the fermentation of food products, which would benefit both the producer and consumer.

2. Contributions of fermented foods to human nutrition

Human health is one of the main reasons behind food choices and this has led to a diverse range of food formulations with specific functionalities that provide better health and wellness. One of the common health disorders associated with diet patterns is gastrointestinal (GI) disorders. Such GI disorders can be prevented to some extent through routine consumption of foods with specific functionality [21]. Hence, the concept of functional foods evolved as the role of food in maintaining health and well-being and therefore gained greater scientific and commercial interest [22]. Lactic acid bacteria and bifidobacteria are well known for their extensive use in the preparation of functional food products [21]. These organisms have been termed as "probiotic bacteria," which does impart certain specific health-promoting attributes through oral feeding. Simultaneously with probiotics, the other term "prebiotics" are known to be non-digestible food ingredients (higher polysaccharides) that have a beneficial effect on the host by selectively stimulating the growth and/or activity of a selected group of bacterial genera and species that are normal colon inhabitants [23].

Probiotics should have the competence to grow and sustain in the human gut in order to deliver health benefits to the host. Therefore, they must have the characteristics to survive while passing *via* the gastrointestinal tract (GIT). During this process, they get exposed to the acidic environment in the stomach, while the bile acid is in the small intestine [24]. Most of the probiotic strains are natural inhabitants of the human intestine and are generally regarded as safe (GRAS) along with acid and bile tolerance and the ability to adhere to gut epithelial cells [25, 26]. Hence, the best designed route for the entry of these probiotic bacteria is the diet, both for animals and human beings [27]. Fermented foods based on milk, cereals, and legumes are among the most accepted food carriers for the delivery of viable probiotic cultures [24]. Probiotic bacteria and their fermentation products appear to influence human health, wherein they provide colonization resistance against potentially pathogenic microorganisms [28].

Fermentation plays mainly important roles like improving the nutritional qualities of food by enhancing the flavor, aroma, and texture of food, contributing toward the preservation of food by the production of main compounds such as lactic acid, acetic acid, alcohol, and alkaline contents. The protein, vitamins, essential amino acids, and fatty acids are enriched by converting the food substrates naturally. During fermentation processes detoxify the food products. Finally, it can be claimed that the fermentation process decreases the cooking periods and requirements of fuel [2].

In several studies, researchers have demonstrated the ability of probiotic bacteria to inhibit pathogenic bacteria by the production of organic acids like lactic and acetic acids during the fermentation process, which lowers the pH of the intestine and consequently inhibits the growth of the undesirable bacteria [29]. In addition to these beneficial health effects, researchers have demonstrated that the major end products of fermentation in humans are short-chain fatty acids (SCFA) like those of acetate, propionate, and butyrate [30, 31]. Besides, a few other antimicrobial substances produced widely by lactic acid bacteria include hydrogen peroxide, carbon dioxide, diacetyl, and bacteriocins [32]. Probiotic bacteria like LAB and bifidobacteria are also known to synthesize folate, vitamin $B_{12,}$ and vitamin K, which are vital components of the human diet and involved in the biosynthesis of nucleotides and cofactors in many metabolic reactions [33].

There has been substantial evidence for the benefits of probiotics and prebiotics in the lowering of (i) lactose intolerance through the activity of β-galactosidase; (ii) antibiotic-associated diarrhea; (iii) colon carcinogenesis; (iv) hypocholesterolemic effect, and (v) gut mucosal dysfunction [34–38].

3. Traditional fermented foods

3.1 Background scenario

Fermentation is one of the oldest methods of preserving food, which became popular at the beginning of civilization as it led to the development of a variety of tastes, flavors, textures, forms, and other sensory attributes. As a process, it involves the transformation of simple raw materials into a range of value-added products, using the phenomena of microorganism growth and their activities on different substrates. Therefore, knowledge of microorganisms' characteristics is essential to understand the fermentation process [4]. Indian-fermented foods such as *dahi*, *rai*, *gundruk*, *koozh*, *kanjika*, *sinki*, *iromba*, *handua*, and *iniziangsang* have been considered as important nutritious diets with significant medicinal properties. Among these fermented food products *koozh*, *dahi*, and *kanjika* are habitually consumed by the local population because of their health benefits [39]. Currently, various fermented foods and beverages have evolved over the years.

3.2 Nutritional status of Indian traditional foods

In the background of a diverse range of traditional Indian foods, the most popular and widely consumed are those based on either milk alone or cereals and legumes with milk. The scientific knowledge on nutritional benefits derived from milk and dairy products is well documented. On the other hand, the same is not true with those of cereals and legumes-based foods, as the raw materials available are specific to the region and store house of complex nutrients. This complexity is linked to the type of fermentation process, product preparation parameters, and final profile that offers ample opportunities to highlight the importance of nutritional constituents in cereals and legumes-based Indian traditional foods.

Cereals and legumes are considered one of the most important sources of dietary proteins, carbohydrates, vitamins, minerals, and fiber for human nutrition. Often, the nutritional quality of cereals and legumes is not on par with that of milk and dairy products. This is further attributed to the complex nature of macronutrients, as well as the prevalence of antinutritional factors, which makes their bioavailability more difficult [40]. In addition, processing methods like soaking, sprouting, milling, fermentation, and cooking/heating have enabled the improvement of nutritional attributes of cereals and legumes [41].

In general, the natural fermentation of cereals and legumes leads to a decrease in the level of complex carbohydrates such as non-digestible poly- and oligosaccharides. In addition, certain amino acids and vitamins, especially B-group vitamins, can be synthesized and become available. Increased amounts of riboflavin, thiamine, niacin, and lysine due to the action of LAB in fermented cereal mixtures were reported in some of the studies [42, 43]. Fermentation also provides ideal pH conditions for the enzymatic degradation of phytate, which is present in cereals in the form of complexes with iron, zinc, calcium, magnesium, and proteins. This reduction in phytate can increase the bioavailability of iron, zinc, and calcium in several folds [40, 44–48]. Thus, fermentation is known to prolong shelf life and impart improved and acceptable texture, taste, and aroma to the final product. During the cereal fermentation process, several volatile compounds are formed, which contribute to a complex mixture of flavors in the products [40]. The presence of aromas representative, that is, diacetyl, acetic acid, and butyric acid make fermented products based on cereals and legumes more appetizing. In many parts of the world, traditional fermented foods are prepared from the most common types of cereals and pulses, such as rice, wheat, soy bean, sorghum. Some of them are utilized as breakfast or light meals, beverages, and colorants, while some are used as main food meals in the regular diet.

Most of these food products are naturally fermented, which mainly encompasses mixed cultures of LAB, yeasts, and fungi. Often, the predominant microflora can be functionally similar, while few others can become functional in a sequential manner with an altered environment due to the fermentation process. Common fermenting bacteria are species of *Lactobacillus, Leuconostoc, Streptococcus, Pediococcus, Micrococcus,* and *Bacillus*. The fungal genera known to be commonly present are *Aspergillus, Cladosporium, Fusarium, Penicillium,* and *Trichothecium*. Common fermenting yeasts are species of *Saccharomyces*, which generally predominate in an alcoholic fermentation [49]. The type of bacterial flora developed in each fermented food depends on the water activity, pH, salt concentration, temperature, and the composition of the food matrix.

The lactic fermentation process enhances the nutritional value, shelf life, safety, and acceptability of a varied range of cereal-based foods [50]. In this process, cereal grains are commonly cleaned, followed by immersion in water for a few days. During this period, there will be the progression of naturally occurring microorganisms which will result in an increase in the LAB population. Moreover, in this course, the endogenous amylases of the grains become active and produce fermentable sugars that aid as an energy source for the LAB. Considering that other practices including grinding, salting, or heating can also change the properties of the final product [47].

A variety of indigenous fermented foods that are prepared in India are basically using cereals with a combination of legumes, which improves the overall protein quality of the food products. Because, cereals contain the highest amount of cysteine and methionine, but lack lysine content, whereas legumes are rich in lysine but lack sulfur-containing amino acids. Therefore, the overall quality of the protein in food products can be improved by mixing both cereals and legumes [51].

In fermented foods like *idli*, *dosa,* and *dhokla*, the fermenting desirable microflora is considered essential for the leavening of the batter and for acid production in *idli* [52, 53]. Fermentation of *idli* batter appears to significantly increase the essential amino acids and simultaneously reduce the antinutrients (phytic acid), enzyme inhibitors, and flatus-causing sugars [54].

3.3 Milk-based fermented foods

Fermentation of milk, either knowingly or unknowingly, has occurred since the earliest times, resulting in various fermented milk products. Fermented dairy products are known for their taste, nutritive value, and therapeutic properties. The nature of these products differs from region to region depending on the indigenous microflora, which in turn depends upon the surrounding environmental factors [55]. The most popular traditional fermented milk products from the Indian subcontinent are *Dahi*, *Cheese* (*Chhurpi*), *Chhur churpen, Philu, mohi*, *Mishti doi*, *Shyow, Somar, Lassi*, *Shrikhand,* and others. The use of desired microorganisms as in the case of controlled fermentation would greatly increase the chances of obtaining products with uniform and consistent quality of acceptable attributes [56].

Lactic acid bacteria convert the lactose from the milk into lactic acid and selective strains produce an antibacterial substance, that is, bacteriocin to destroy milk curdling by unwanted bacteria. For example, in milk, when *Lactococcus lactis* is inoculated, it converts lactose into energy (Adenosine triphosphate ATP) by enzyme synthesis. Here, lactic acid is the byproduct of ATP that coagulates milk, which can be utilized for the preparation of whey and cheese. Therefore, the bacterium is not just curdling the milk, but it preserves the products by inhibiting the growth of unwanted microorganisms that is by lowering the pH of the product by lactic acid production.

Some "food grade" starter strains, that is, *L. lactis* ssp. *lactis* produce nisin, are an antibiotic-like substance called a bacteriocin. It has natural antimicrobial activity against a wide variety of Gram-positive bacteria, including food-borne pathogens such as *Listeria monocytogenes, Staphylococcus aureus, Bacillus cereus,* and *Clostridium* sp. It is believed that the nisin's primary target is the cell membrane, which does not need a receptor to interact with the cell membrane of the microbe. It is a natural preservative prevalent in cheeses prepared with *L. lactis* ssp. *lactis* and is currently recognized as a safe food preservative. Hence, it is used as a preservative in thermally processed and low pH foods, also in various pasteurized dairy products and canned vegetables, baked, beverages, high-moisture flour products, and pasteurized liquid eggs. As nisin cannot be chemically synthesized, the nisin-producing *L. lactis* strains are used for their industrial synthesis. A highly purified nisin preparation has led to an interest in the use of nisin for human ulcer therapy and mastitis control in cattle, while different enzymes and other metabolic products produced by *Lactococcus lactis* contribute to the more subtle aromas and flavors that distinguish different cheeses.

Most foods mentioned below in this category are prepared by simply adding LAB to the milk of cow, buffalo, or yak and allowed to ferment.

3.3.1 Dahi

Dahi is one of the most popular fermented dairy products on the Indian subcontinent highly accepted for its mild acidic taste and pleasant flavor. The main properties of good quality *dahi* are steady with uniform consistency, sweet aroma, and light sour taste. The surface of the *dahi* will be smooth, shiny, and free of cracks as well as gas bubbles. It is believed that *dahi* possesses nutritional and therapeutic values at a higher level when compared to the milk utilized for its preparation. *Dahi* is easy to digest and has been found to possess several health benefits [57]. One of the primary compositions of *dahi* made with whole milk is as follows: water 85–88%, fat 5–8%, protein 3.2–3.4%, lactose 4.6–5.2%, lactic acid 0.5–1.1%, ash 0.7–0.75%, calcium 0.12–0.14%, and phosphorus 0.09–0.11% as reported by Laxminarayana et al. [58].

The lactic acid bacterial cultures commonly associated with the inoculum are strains of *Lactococcus lactis* ssp. *lactis*, *Lactococcus lactis* ssp. *cremoris*, *Lc. lactis* ssp. *diacetylactis*, *Leuconostoc cremoris*, *Lactobacillus delbrueckii* ssp. *bulgaricus*, *Lactobacillus acidophilus*, and *Lactobacillus helveticus*. Choosing a good starter culture is important to attain a good flavor with desirable characteristics. Both mesophilic and thermophilic starters have been reported to be used in a number of different combinations [59, 60]. The addition of probiotic cultures such as *Lactobacillus* acidophilus and Bifidobacterium bifidum combined with the regular lactic cultures for *dahi* preparation helps to increase the therapeutic and nutritional value [61]. *Dahi* has been recognized as a potential source of lactic acid bacteria, chiefly species of *Lactobacillus* that are active against foodborne pathogenic and spoilage bacteria such as *Staphylococcus aureus*, *Bacillus cereus*, Bacillus licheniformis, Bacillus subtilis, Escherichia coli and *Klebsiella* sp., and *Pseudomonas* sp. [62]. In addition, a strain of *Lactobacillus* delbrueckii ssp. *bulgaricus* produces heat-resistant bacteriocin with broad-spectrum antibacterial activity and the potential for use as a food biopreservative isolated from *dahi* [63]. As such, *dahi* differs from yogurt in the use of mixed starters of mesophilic lactococci. Diacetyl is one of the main metabolites that induce flavor in *dahi* products, which is accepted more by South Asian origin people than acetaldehyde flavor in yogurt (**Figure 1**) [64].

Figure 1.
Dahi (courtesy: Zee news).

3.3.2 Cheese

Cheese (*Chhurpi*) made from yak and cattle milk is mainly consumed in the Himalayan plateau and industrial production of this product has not yet become the norm. *Chhurpi* is in white color, soft in texture with a mild to strong flavored taste, and is eaten as a curry blend with edible wild ferns (*Diplazium* spp.), pickles, and condiments along with cooked rice in meals [65]. Yak cheese contains the following chemical composition, 68.2% of total solid, 49.4% of butterfat on a dry matter basis, and 1.37% of salt [66, 67]. In mature *Chhurpi*, LAB count 7.5 log CFU/g was recorded. All LAB strains except *Leuconostoc mesenteroides* BFE1637 showed a high level of hydrophobicity. This is an important property of LAB that helps to colonize epithelial cells. LAB strains produce enzymes such as peptidases and esterase-lipases that play an important role in the improvement of *cheese* quality.

Similarly, *Chhu*, *Shyow*, *Mohi*, *Somar*, and *Philu* are traditional fermented milk products in the Himalayan plateau, predominantly fermented by LAB. *Chhu* (*Sheden*) is a strong-flavored cheese-like product, LAB were mainly present at 8.1–8.8 log CFU/g. *Shyow* is a thick curd-like product, prepared with yak milk. *Mohi* is buttermilk, prepared with churning *dahi*, consumed as a refreshing beverage. *Somar* is a soft paste, strong flavored with a bitter taste, eaten as a soup along with cooked rice or finger-millets, whereas *Philu* is a typical indigenous cream-like milk product obtained from cow milk or yak milk and is consumed as a cooked paste delicacy with boiled rice [68, 69]. These products are rich in LAB that produces various enzymes such as esterase, phosphatase, leucine-arylamidase, β-galactosidase, and peptidase. These bacterial strains inhibited the growth of pathogens such as Enterobacter agglomerans, Enterobacter cloacae, and Klebsiella pneumoniae. Fermented milk products have been reported to contain more common LAB such as Streptococcus cremoris, Streptococcus lactis, Streptococcus thermophilus, *Lactobacillus* bulgaricus, *Lactobacillus* acidophilus, *Lactobacillus* helveticus, *Lactobacillus cremoris, Lactobacillus* plantarum, *Lactobacillus* curvatus, *Lactobacillus* fermentum, *Lactobacillus* paracasei *subsp. pseudoplantarum, Lactobacillus* alimentarius, *Lactobacillus kefir, Lactobacillus* hilgardii, Enterococcus faecium, *L. mesenteroides, Lactobacillus* farciminis, *Lactobacillus* brevis, Lactococcus lactis *subsp. cremoris, Lactobacillus* casei *subsp. Casei*, and *Lactobacillus* bifermentans (**Figure 2**) [39].

Figure 2.
Cheese (courtesy: Healthy nibbles).

3.3.3 Mishti doi

Mishti doi also called as *lal dahi* or *payodhi* is most popular in eastern India. It is a blend of sweet *dahi* that usually appears as a light brown color and has a firm texture with a cooked or caramelized flavor. There are two types of formulations have been recognized for the preparation of *Mishti doi* [70]. One combination involves the use of *Streptococcus salivarius* ssp. *thermophilus, Lactobacillus acidophilus* and *Lactobacillus delbrueckii* ssp. *Bulgaricus,* and the other formula are on *Lactobacillus* acidophilus, Lactococcus lactis ssp. *Lactis*, and *Saccharomyces cerevisiae*. Microbiological study of market samples of *mishti doi* showed the presence of yeast such as *Saccharomyces*, *Candida,* and *Rhodotorula*, as well as the LAB such as *Lactobacillus, Lactococcus,* and *Streptococcus*. Protease peptone, a trace protein of milk, plays a key role in imparting the brown color to the product (**Figure 3**) [71].

3.3.4 Lassi

Lassi also known as buttermilk from an Indian perspective and is one of the popular lactic-fermented milk-based beverages in the Indian subcontinent, which is consumed predominantly during hot and warm seasons. It is a by-product obtained during the desi butter preparation from *dahi* by churning. It is also made by crushing

Figure 3.
Mishti doi (courtesy: Babs projects).

the set *dahi* with the agitator and adding the required amount of water, sugar, or salt and flavor compounds. Rangappa and Achayya in 1974 [72] stated that the composition of each *Lassi* differs mainly due to the variety of milk used, the amount of dilution made while churning, and the efficiency of fat removal. A typical composition of *Lassi* includes water 96.2%, fat 0.8%, protein 1.29%, lactose 1.2%, lactic acid 0.44%, ash 0.4%, calcium 0.6%, and phosphorus 0.04%. To extend the shelf life of *Lassi* beyond 6 days 0.03–0.35% sodium metabisulfite must be added and stored at 37°C. The sulfur flavor developed by the preservative can be masked by adding 0.07–0.09% crushed ginger and 0.5–0.7% salt [73]. The preservative effect of nisin in improving the shelf life of *Lassi* has been studied, wherein *Lassi* could be stored for 32–48 h at 30°C with the addition of Nisaplin at a concentration of 200–500 IU/ml, while the stability was 8–10 days when stored in a refrigerator [74]. In order to increase the therapeutic value of *Lassi*, the method of preparation has been standardized using probiotic culture of *Lactobacillus* acidophilus along with Streptococcus thermophilus to obtain a desirable flavor in the final product (**Figure 4**) [75].

3.3.5 Shrikhand

Shrikhand is a sweetened fermented traditional dairy product extensively consumed in the western and northern parts of India. It has a refreshing taste with a pleasant aroma, smooth and homogenous texture, and firm consistency. The preparation of *Shrikhand* includes curd products (*dahi*) made by lactic fermentation of complete milk, either cow or buffalo milk, after which whey is removed from the curd using a hanging muslin cloth bag hung for 6–8 h. The resultant solid mass (called chakka) is evenly mixed with ground sugar (44–45%) and prepared as a semi-solid mass to which flavorings such as cardamom and saffron are mixed. The cultures used are mesophilic LAB, the same cultures involved in the preparation of *dahi* [76].

The technological and microbiological aspects of the *Shrikhand* preparation are reported in an earlier study [77]. Preparation of nutritional *Shrikhand* using buffalo skimmed milk fermented with a combined culture of 2% Lacto*bacillus acidophilus* (NDRI-AH1) and Str*eptococcus salivarius* ssp. *thermophilus* (NDRI-YHS) has been

Figure 4.
Lassi (courtesy: Swasthi's recipes).

Figure 5.
Shrikhand (courtesy: Babs projects).

reported to reduce the high-fat content in the final product [78]. A change in the mineral content profile of milk for the final *Shrikhand* product was studied [79]. Post-production heat treatment (PPHT) of *Shrikhand* at 70°C for 5 min improved the shelf life up to 15 days at 35°C and up to 70 days at 8–10°C [80, 81]. *Shrikhand* powder has been developed by spray drying method that is by exposing the product at 160–170°C of inlet temperature and 100°C of outlet temperature. Dehydrated *Shrikhand* would have a shelf life of 90 days when stored in gas-packed containers at 30°C (**Figure 5**) [82, 83].

3.4 Cereal and legume-based fermented foods

Cereal and legume-based fermented foods are regarded as staple foods in their respective provinces. They are a major source of economical dietary energy and nutrients throughout the world. In Indian subcontinent, region-specific cereals or legumes are subjected to natural or controlled fermentation to obtain desirable end products through the involvement of desirable microorganisms, mainly LAB, yeasts, and fungi, and have been well documented [84–86], and these organisms have the ability to increase palatability, maintaining the quality, safety, and nutritional value of the raw materials. The successive phase of growth of microorganisms in the fermentation of cereals and legumes also favors the growth of yeasts, which often occurs as a component of mixed microflora and gives specific characteristics to the product [87].

Most foods such as *idli, dosa, dhokla, khadi, Punjabi warri, adai dosa, kallappam, ambali* or *pazhaiya soru, koozhu, nan,* and *parotta* are routine food products of the native population. For the preparation of this type of fermented food products, chiefly use cereals such as rice (*Oryza sativum*), wheat (*Triticum* spp.), ragi (*Eleusine coracana*), barley (*Hordeum vulgare*), whereas pulses include black gram, green gram, and red gram. These cereals and legumes are cultivated in India since the Indus valley civilization period, that is, 9000–5500 BC [88], and are commonly used as main components in the preparation of significant amounts of food products. They are one of the effective substrates for the preparation of functional foods combined with probiotics. Since these products are rich in non-digestible carbohydrates, which help in the growth of Lactobacilli and Bifidobacteria. They consist of water-soluble fibers such as β-glucan, galactooligosaccharides, fructooligosaccharides, and arabinoxylan, and these fibers can be digested by selective LAB strains [89].

In the preparation of fermented foods such as *idli, dosa, adai dosa, kallappam,* and *dhokla,* the batter is prepared from the basic ingredients such as milled rice (*Oryza sativus*) and dehulled Black gram (Phaseolus mungo), and left the batter to ferment overnight at room temperature. For this, sodium bicarbonate can be added to create anaerobic conditions for LAB and yeast growth. In case of *Kallappam* preparation, the batter can be supplemented with fermented toddy to provide the extra LAB source. Leuconostoc mesenteroides are the most commonly encountered bacterium [90]. *Lactobacillus* plantarum AS1 isolated from fermented food from Southern Indian *Kallappam* successfully prevented colonization of entero-virulent bacterium Vibrio parahaemolyticus in HT-29 cell line and colorectal cancer in male Wistar rats [91, 92].

Preparation of *Pazhaiya soru* or *ambali* (Fermented rice) involves adding water to the cooked rice and later incubating the mixture overnight. Before consuming buttermilk add salt directly [20]. In most parts of south India, farmers consume this as an early morning meal before heading to the farm-field. Major microbiota isolated from this type of food include *Lactobacillus plantarum, Lactobacillus fermentum, Leuconostoc mesenteroides, Pediococcus cerevisiae,* Pediococcus acidilactici, Enterococcus faecalis, and Weissella paramesenteroides.

Koozhu is a traditional South Indian fermented food, a type of porridge made from finger millet and claimed to be a nutritious food. It is included in the daily diet of rural agricultural workers and urban households [93]. It is also given to children at weaning age [94]. It is made from Kezhvaragu or Cumbu flour and broken rice in a clay pot. *Koozhu* is easily digested and cools the body; therefore, during the summer, street vendors sell it as a cool drink in Southern India. There is an increase in thiamine, riboflavin, and niacin contents during its fermentation [95, 96].

In India, there are a large proportion of traditional fermented foods that are still unexplored for the microbiota and therapeutic values. Most of the research work has been done on the following fermented food products.

3.4.1 Idli

Among the closely related types of traditional fermented foods based on the combination of cereal and legume is the *idli*. It is a white, fermented (acid-fermented), steamed product with a soft and spongy texture, more popular and consumed throughout Southern India. It has been documented that *idli* batter fermentation has been in use since 1100 AD [97]. It is the resulting product of naturally fermented batter made from washed and soaked milled rice (*Oryza sativus*) and dehulled Black gram (Phaseolus mungo). From a nutritional and health point of view, *idli* seems to be an ideal human food for all ages of people and at all times. Investigations into the primary aspects of *idli* batter fermentation were started as early as in 1955 at the Central Food Technological Research Institute, Mysore, India. Several researchers have used different proportions of Black gram cotyledons to rice ranging from 4:1 to 1:4 weight to weight (w/w) for making *idli* with a preference for 2:1 and 3:1 over 4:1 [77, 98–101].

Studies have demonstrated the optimum fermentation conditions for obtaining good *idli*s as well as the physiochemical and microbiological changes that occur during intermittent periods of incubation at varying temperatures [98, 102, 103]. Typically, the microorganisms that develop during the initial and subsequent soaking of the ingredients are sufficient to cause fermentation. The microbiological changes during the fermentation period have shown the involvement of different genera and species of LAB and yeasts. The main bacterial floras identified include *Lactobacillus* brevis, Leu*conostoc mesenteroides, Lactobacillus* delbrueckii, *Lactobacillus* lactis, *Lb.*

fermentum, Pediococcus cerevisiae, and Str*eptococcus* faecalis, while the yeast flora comprised Torulopsis holmii, Torulopsis candida, *Candida kefyr, Candida cacaoi,* Can*dida* tropicalis, *Candida fragicola,* Hansenula anomala, and *Puccinia* graminis. Moreover, these studies have shown the presence of major microflora at different stages of *idli* fermentation [101, 102, 104–108].

Two important changes that occur during *idli* fermentation are acidification and the leavening of the batter. Comprehensive studies on the various changes that accompany *idli* batter fermentation have shown that in addition to a consistent increase in microbial populations, the pH dropped to 4.4–4.9 from an initial pH of 6.6 [102, 106]. Several attempts have been made to improve holding *idli* fermentation by standardizing various physicochemical factors. An increase in the fermentation rate of *idli* batter was observed to accompany a rise in temperature [98]. Fortification of *idli* batter with glucose at 1% level has shown a beneficial effect on the gas formation and leavening during fermentation [109]. Similarly, the relationship between microflora type and biochemical traits revealed an increase in the content of water-soluble group B vitamins during *idli* fermentation [110].

As a step toward convenience in preparation, the dry mix concept was proposed as early as in 1960 [98]. In a similar approach, a process related to an improved means of providing inoculum (LAB and yeast) in ready-to-use form *idli* fermentation was developed [111]. In order to reduce the fermentation time of *idli* batter and increase its shelf life of fermented *idli* batter, an Indian patent has been filed [112]. Simultaneously, the flavor profile of such controlled fermented *idli* batter has shown the presence of desirable flavor compounds such as ketones, diols, and acids for an 8-day storage period. This flavor profile can be a reliable qualitative and quantitative parameter for objective evaluation [113].

Plantaricin LP84, a bacteriocin produced by *Lactobacillus* planatrum NCIM 2084, was able to retard the growth of foodborne pathogens such as *Bacillus cereus* F 4810, *Escherichia coli* D 21, and Staph*ylococcus* aureus FRI 722 during *idli* batter fermentation [114]. Fermented *idli* is easy to digest and is often used as baby food. This is the prescribed diet in hospitals for patients undergoing treatment (**Figure 6**) [85].

3.4.2 Dosa

Dosa is a fermented, thin, crispy, baked, and pancake-like product widely consumed in southern and parts of western India. In recent years, *dosa* has become more

Figure 6.
Idli and batter (courtesy: Sharmis passions, Hebbars kitchen).

Figure 7.
Dosa (courtesy: Madhura's recipe).

prominent throughout India including in Indian restaurants around the world. The *dosa* batter preparation is nearly similar to the *idli* batter, except for the proportion of milled rice and Black gram dhal usage. In the preparation of *dosa* batter, an equal quantity of milled rice and dehulled black gram dhal is soaked in water for a period of 6–8 h at room temperature (25–30°C) and grinded into a fine paste using the required amount of water, that is, 2.0–2.5 parts by weight. Later allow the batter to undergo natural fermentation for a period of 10–12 h at room temperature (25–30°C). From this fermented batter, *dosa* can be made like thin, crispy, pancake-type product.

Even in fermented *dosa* batter also numerous biochemical changes occur due to the effect of lactic acid bacteria along with yeasts [115, 116]. The predominant species identified were *Leuconostoc mesenteroides, Lactobacillus fermentum, Lactobacillus delbrueckii, Pediococcus cerevisiae, Saccharomyces cerevisiae*, *Hansenula anomala*, and *Kluyveromyces* sp. Typically, these microorganisms come from raw materials such as rice and Black gram [106]. *Dosa* batter fermentation also resulted in increased biochemical attributes, including that of water-soluble group B vitamins such as thiamine, riboflavin, and cyanocobalamin.

Attempts have been made to prepare products similar to *dosa* by replacing Black gram with other legumes like soy beans. Soy bean-based *dosa* batter was found to be nutritious but less preferred organoleptically (**Figure** 7) [117].

3.4.3 Dhokla

Dhokla is one of the famous steamed food products in the western part of India, especially in Gujarat and Maharashtra States. It is prepared with a mixture of Bengal gram (Cicer arietinum), dehulled black gram (Phaseolus mungo), and milled rice (Oryza sativa) in a ratio of 2:1:1. This composition gives the product a soft and spongy texture. The above-mentioned mixture of grains is soaked in water for 6 to 8 h and ground to a granular consistency. To the resultant batter, the curd is added to a proportion of 1:1.5 w/w, after which it is allowed to ferment for 16 to 18 h and then steamed the product. Usually, this product is consumed by seasoning with oil, spices, and coriander leaves for taste [85, 97].

Research studies have shown that the numbers of LAB and yeast cells are increased during the fermentation process. Major microorganisms such as *Lactobacillus fermentum*, *Leuconostoc mesenteroides*, and *Han. Silvicola* are present in this product, whereas biochemical changes such as pH, acidity, and volatile fatty acid content in the food product are also documented as influencing the increase in microbial counts [118].

Figure 8.
Dhokla (courtesy: Ram Asrey).

One more study found that *dhokla* batter was prepared using *Lactobacillus* species that exhibited antibacterial activity against harmful microorganisms such as *Bacillus* subtilis, Ba*cillus* licheniformis, and *Brevibacillus* laterosporus present as contaminants before processing. But in another study, it was reported that the same harmful bacterial species grew well in the *dhokla* batter that was prepared using *Lactococcus* species due to a lack of antibacterial activity [119, 120]. These results indicate that those LAB have an antagonistic nature and are able to provide healthy food products in terms of safety (**Figure 8**).

3.4.4 Kadhi

It is a traditional fermented food, prepared by boiling lactic fermented and agitated *dahi* with 5 to 8% (w/w) Bengal gram flour, that is, besan flour as a thickening agent. It is consumed in most parts of India as a cooked food that has a slightly sour taste and gives it a classic baked flavor. Considering the potential use of antagonistic LAB, studies have been carried out to evaluate the effectiveness of the antibacterial properties of selected LAB for foodborne pathogenic and spoilage bacterial species occurring in pre-processing and post-processing contaminants in *kadhi*. The study highlighted the benefits of using pure cultures of LAB with antagonistic nature such as *Lactobacillus delbrueckii* ssp. *lactis* CFR 2023 and *Lactobacillus* delbrueckii ssp. *bulgaricus* CFR 2028 in *kadhi* preparation with desirable quality attributes and preservation against foodborne pathogens [63, 120].

LAB isolated from *khadi* showed antibacterial activity against *Staphylococcus* aureus, Escherichia coli, and *Pseudomonas* aeruginosa as well as exhibited probiotic properties [121]. The effectiveness of antagonistic cultures of *Lactobacillus* species was evaluated in terms of microbiological and sensory parameters in *kadhi* during storage at ambient and refrigerated temperatures [119]. Research studies have attempted to modify the *kadhi* preparation to provide better nutritional status through the use of smaller quantities of Bengal gram flour obtained from pre-soaked seeds. Based on the sensory attributes and the characteristic consistency of *kadhi*, the product prepared using Bengal gram flour (from pre-soaked seeds at 40 mg/g) showed better acceptance compared to the product prepared by a conventional method using 100 mg/g of besan (**Figure 9**) [56].

3.4.5 Punjabi warri

It is a pulse-based dried product, commonly prepared in north India. They are spicy, hollow, crunchy, small fried balls and used as a condiment in cooking with vegetables, soups, or Indian sambhar (dhal-based spicy liquid). It is prepared using dehulled black gram (Phaseolus mungo), which needs to be soaked overnight in water

Figure 9.
Kadhi (courtesy: We recipes).

and then ground into a soft batter. Different types of spices are added to this batter and made into small balls. After this allowed to ferment for a few hours and dried in an open space for 4 to 10 days [122].

Microbiological and biochemical aspects associated with Punjabi warri have been studied in some research investigations [106, 122]. These studies have established that the development and prevalence of microorganisms were affected by the seasons; summers are more favorable for bacteria and winters for yeasts. These microbial types tend to increase significantly as fermentation progresses. The microorganisms that are responsible for the fermentation of this product mainly include Leu*conostoc* mesenteroides, *Lactobacillus* fermentum, Str*eptococcus* faecalis, *Saccharomyces cerevisiae*, Pichia membranaefaciens, and *Trichosporon beigelii*. During this process, there is an increase in enzyme activity such as amylase and other proteinase, which leads to the enhancement of water-soluble B-vitamins such as thiamine, riboflavin, and cyanocobalamin. These biological changes influence the nutritional quality of food products.

Some research studies have shown that the use of *Lactobacillus* delbrueckii ssp. *bulgaricus* CFR 2028 and *Lactobacillus* delbrueckii ssp. *lactis* CFR 2023 in warri preparations exhibited antibacterial activity against food-borne pathogens like B*revibacillus* laterosporus, Ba*cillus* licheniformis, and *Bacillus subtilis*. The preparation of this product involves combined fermentation and drying processes where *Lactobacillus* species act as biopreservative. Hence, pathogenic microbial growth was stopped while the storage period of 10 days at room temperature [119, 120].

In addition to the above, there are several other documented cereal and legume-based traditional foods popular in specific regions of India. However, detailed scientific and technological studies have not been much documented on these foods.

3.5 Milk and cereal/legume-based fermented foods

3.5.1 Rabadi

Rabadi is a fermented drink most commonly used in the Rajasthan province of India. It is made from a combination of pearl millet flour (Pennisetum typhoideum) or wheat flour and buttermilk, using an earthenware vessel as a container and then keeping it for natural fermentation for 4 to 6 h at room temperature. This is followed by dilution of the product with water, cooking, and adding salt to taste. This process improves the level of LAB that can be served as health drinks [123].

Rabadi fermentation of freshly ground wheat millet flour brought about a significant increase in the HCl-extraction capacity of calcium, iron, copper, zinc, manganese, and phosphorus [124]. Consumption of such fermented foods may help to improve prevailing mineral deficiencies due to their limited bioavailability of such coarse grains [125–128]. The effect of processing parameters such as dehulling, cooking, and fermentation on the antioxidants present in pearl millets during the preparation of *rabadi* revealed that cooking and fermentation result in improved flavonoids [129]. Furthermore, using germinated pearl millet grains, the optimization of *rabadi* preparation was performed by response surface methodology. The most acceptable product is prepared with 5.3% flour and 72% water based on the type of curd (**Figure 10**) [11, 130].

3.5.2 Kulcha

Kulcha is a popular product consumed in north India, which is lately gaining popularity in other provinces of India. It is prepared by mixing the main components such as white wheat flour, milk, sugar, salt, curd, dry yeast, baking powder, and water. It is thoroughly mixed and kneaded into firm dough. The dough is allowed to ferment naturally for a period of 6 to 8 h at room temperature (28 to 30°C). A small ball of dough of uniform size is used to make the thick disc-shaped *kulchas* by hand and then a flat dough is baked in a tandoor (a metallic baking sheet or special oven made of clay) to a golden brown color and served hot (**Figure 11**) [56].

3.5.3 Naan

Naan is a fermented flatbread baked in a clay oven called a tandoor and is widely consumed by people in northern India. In recent times, *naan* is becoming more

Figure 10.
Rabadi (courtesy: Dishes guru).

Figure 11.
Kulcha (courtesy: Times food).

Figure 12.
Naan (courtesy: Recipe pocket).

popular in other provinces of India. To make *naan*, the first dough is made using components such as white wheat flour, egg, milk, curd, baking powder, salt, and sugar. This is allowed for fermentation at room temperature for 1 to 2 h. A small portion of the dough is taken that is rolled out on a flat surface and roasted in a clay oven until it turns brown and crispy on both sides. Finally, it is served with butter (**Figure 12**) [56].

3.6 Vegetable-based fermented foods

The ancient civilization was well aware of the existence of natural microflora and its role in the fermentation of vegetables, which could result in palatable foods for human consumption. Vegetables contain low sugar, neutral pH, and their composition is not favorable to the spontaneous growth of LAB. However, over the centuries, people have traditionally developed methods of lactic fermentation that could stabilize and improve the nutritional quality of vegetables. Fermented vegetables represent an essential element of the human diet. Lactic acid fermentation, which improves the organoleptic and nutritional quality of the vegetables, has remained more of a domestic- or cottage-level process.

Spices and herbs impart a fine flavor and play a key role in fermented vegetables. Spices such as garlic, clove, and chili inhibit the growth of food-borne microorganisms because these spices contain antagonistic activity. Some aromatic compounds such as terpenes and polyphenols (found in spices), allyl isothiocyanate (found in mustard seed) as well as sulfur (found in garlic) have antimicrobial activity and selectively stimulate the growth of LAB. Mustard seed oil is most commonly used in north India since it has the property to promote the lactic fermentation of food products, which helps in their long storage. Apart from this, chemical preservatives such as sorbic and benzoic acids are used in the development of vegetable-based fermented products [131]. The concentration of salt induces plasmolysis in vegetables, thus promoting anaerobiosis for the proliferation of lactic acid bacteria. Some of the well-known vegetable-based lactic fermented products are presented in the following paragraphs.

Popular types of pickles consumed by the human population of India have been those based on unripe mangoes, goose berries, lemons, swallow root (*Decalepis hamitonii*), and a variety of mixed vegetables. In general, most of these pickles are prepared at home, subjecting the vegetables to natural fermentation. In a specific type of pickles, preservation is achieved through lactic acid fermentation and in the presence of high concentrations of salt. The process involves washing raw materials, then cutting them into appropriately sized shapes, and mixing them with salt at a level twice the weight of the raw material. Necessary spice powders are also used in the preparation. The powdered spice mix mainly includes chili, mustard, and coriander seeds. The product is allowed to ferment in the closed container for 8 to 10 days at room temperature. The aging of the product gives it a slightly acidic taste and pickles develop acceptable organoleptic properties. Although not much research has been done on the nature of microflora and other attributes, it is believed that the microflora mainly comprises lactic acid bacteria and to some extent acetic acid bacteria.

In another specific type of pickle that is devoid of any liquid, the raw materials for the preparation of the pickle are the same as described above and preservation is achieved through a high concentration of salt, a mixture of spices, and edible oil. The prepared pickle product is intermittently fried with oil for 3 to 4 times and placed in a closed container to ferment. Moreover, the product should be covered with a sufficient quantity of edible oil, where the oil used depends on the specific regions of this country. In this specific type of preparation, fermentation occurs naturally with the predominance of LAB, which can survive and grow in the presence of high concentrations of salt. Regardless of the type of pickle preparation, the shelf life is quite reasonable, extending to periods of 6 months and beyond if proper hygiene and sanitation practices are in place during the preparation and subsequent storage. In the absence of any microbiological studies on the nature of pickle fermentation, from the product profile, it appears that species of *Pediococcus* tend to predominate over other LAB.

There are several other traditional fermented vegetable-based foods, which are more popular in eastern, northern, and north-eastern regions of India. A few of them known by traditional names are *gundruk*, *sinki*, *Iromba*, *rai*, *Kanjika* or *kanji*, and others. They are considered as good appetizers and tribal people use these foods for indigestion therapies [132]. *Gundruk* soup is usually given to breast-feeding mothers to improve milk efficiency. It is considered as a tonic for elderly people [133]. *Sinki* is a fermented radish root that more effectively cures diarrhea and stomach pain. *Iromba* made from tree bean (Parkia roxburghii) is used as a starter [133]. Fermented *rai* has

Figure 13.
Vegetable-based fermented foods.

Figure 14.
Pickles (courtesy: Cookpad.com, times of India).

the therapeutic benefit which resolves stomach pain and gas problems and considerably improves digestion [133]. *Kanjika* or *kanji* is a lactic fermented rice product, recommended for several types of chronic diseases in Indian Ayurvedic medicine [134]. Carrot *Kanji* is known for its high nutritional value which has energizing as well as relaxing properties [135]. Beetroot *kanji* has the property to prevent infectious and malignant diseases (**Figures 13** and **14**) (**Table 1**) [136].

Product	**Region**	**Microorganism(s)**	**Substrate**
Milk-based			
Dahi	India	*Lactococcus lactis* ssp. *lactis*, *Lactococcus lactis* ssp. *cremoris*, *Lc. lactis* ssp. *diacetylactis*, *Leuconostoc cremoris*, *Lactobacillus delbrueckii* ssp. *bulgaricus*, *Lactobacillus acidophilus*, and *Lactobacillus helveticus*	Whole milk [57–64]
Cheese (*Chhurpi*, *Chhu*, *Shyow*, *Mohi*, *Somar*, and *Philu*)	Himalayan plateau	*Streptococcus cremoris*, *Streptococcus lactis*, *Streptococcus thermophilus*, *Lactobacillus bulgaricus*, *Lactobacillus acidophilus*, *Lactobacillus helveticus*, *Lactobacillus cremoris*, *Lactobacillus plantarum*, *Lactobacillus curvatus*, *Lactobacillus fermentum*, *Lactobacillus paracasei subsp. pseudoplantarum*, *Lactobacillus alimentarius*, *Lactobacillus kefir*, *Lactobacillus hilgardii*, *Enterococcus faecium*, *Leuconostoc mesenteroides*, *Lactobacillus farciminis*, *Lactobacillus brevis*, *Lactococcus lactis subsp. cremoris*, *Lactobacillus casei subsp. Casei*, and *Lactobacillus bifermentans*	Yak and cattle milk [39, 64–69]
Mishti doi (*lal dahi* or *payodhi*)	Eastern India	*Streptococcus salivarius* ssp. *thermophilus*, *Lactobacillus acidophilus*, *Lactobacillus delbrueckii* ssp. *Bulgaricus*, *Lactococcus lactis* ssp. *Lactis*, and *Saccharomyces cerevisiae*	Blend of sweet *dahi* [70, 71]
Lassi (Buttermilk)	Northern India	*Lactobacillus acidophilus*, *Streptococcus thermophilus*	By-product of Dahi (Butter) [72–75]
Shrikhand	Western and northern parts of India	*Lactobacillus acidophilus* and *Streptococcus salivarius* ssp. *thermophilus*	Cow or buffalo milk [76–83]
Cereal and legume-based			
Idli	Southern India	*Lactobacillus brevis*, *Leuconostoc mesenteroides*, *Lactobacillus delbrueckii*, *Lactobacillus lactis*, *Lb. fermentum*, *Pediococcus cerevisiae*, and *Streptococcus faecalis*, yeast flora comprised *Torulopsis holmii*, *Torulopsis candida*, *Candida kefyr*, *Candida cacaoi*, *Candida tropicalis*, *Candida fragicola*, *Hansenula anomala*, and *Puccinia graminis*	Milled rice (*Oryza sativus*) and dehulled Black gram (*Phaseolus mungo*) [77, 85, 97–114]

Product	Region	Microorganism(s)	Substrate
Dosa	Southern India	*Leuconostoc mesenteroides, Lactobacillus fermentum, Lactobacillus delbrueckii, Pediococcus cerevisiae, Saccharomyces cerevisiae, Hansenula anomala*, and *Kluyveromyces* sp.	Milled rice (*Oryza sativus*) and dehulled Black gram (*Phaseolus mungo*) [106, 115–117]
Dhokla	Western part of India	*Lactobacillus fermentum, Leuconostoc mesenteroides* and *Han. Silvicola*	Bengal gram (*Cicer arietinum*), dehulled black gram (*Phaseolus mungo*), and milled rice (*Oryza sativa*) [85, 97, 118–120]
Kadhi	Indian subcontinent	*Lactobacillus delbrueckii* ssp. *Lactis,* and *Lactobacillus delbrueckii* ssp. *bulgaricus*	Dahi (Curd) with 5 to 8% (w/w) Bengal gram flour [56, 63, 119–121]
Punjabi warri	Northern India	*Leuconostoc mesenteroides, Lactobacillus fermentum, Streptococcus faecalis, Saccharomyces cerevisiae, Pichia membranaefaciens,* and *Trichosporon beigelii*	Dehulled black gram (*Phaseolus mungo*) and different types of spices [106, 119–122]
Milk and cereal/legume based			
Rabadi	Western part of India	Lactic acid bacteria	Pearl millet flour (*Pennisetum typhoideum*) or wheat flour and buttermilk [11, 123–130]
Kulcha	Northern India	Lactic acid bacteria	White wheat flour, milk, sugar, salt, curd, dry yeast, baking powder, and water [56]
Naan	Northern India	Lactic acid bacteria	White wheat flour, egg, milk, curd, baking powder, salt, and sugar [56]
Vegetable-based [131–136]			
Pickles of unripe mangoes, goose berries, lemons, swallow root (*Decalepis hamitonii*) and a variety of mixed vegetables	Indian subcontinent	Lactic acid bacteria	Different types of vegetables along with spices like garlic, clove, chili, mustard seeds, mustard seed oil/groundnut oil, curry leaves,
Gundruk	North-eastern regions of India	Lactic acid bacteria	Green leaves of mustard, radish, and cauliflower
Sinki	North-eastern regions of India	Lactic acid bacteria	Radish root
Iromba	North-eastern regions of India	Lactic acid bacteria	Tree bean (*Parkia roxburghii*)

Product	Region	Microorganism(s)	Substrate
Rai	North-eastern regions of India	Lactic acid bacteria	Mustard green leaves
Kanjika or kanji	North-eastern regions of India	Lactic acid bacteria	Rice

Table 1.
Examples of traditional fermented foods of India.

4. Concluding remarks

Indigenous fermented foods have played a vital role in the history of human health. They can be produced and distributed at a relatively low cost. They are typically highly nutritious, providing calories, protein, vitamins, and minerals at prices most consumers can afford. As canned and frozen foods are unavailable or too expensive for hundreds of millions of economically deprived populations around the world, traditional fermented foods can fill this gap. Acid fermentation combined with salting remains one of the most practical methods to preserve and increase the organoleptic and nutritional value of fresh vegetables, cereal porridge, and milk-cereal mixtures. Furthermore, ethanol fermentation is very significant in preserving and increasing the nutritional value of cereal grains and fruit juices. Modern food technology exploits enrichment or fortification to improve the nutritional value of foods to reach consumer needs.

Author details

Arekal Nagaraja Roopashri[1*], Janakiraman Savitha[1], M.S. Divyashree[2], B.S. Mamatha[3], K.A. Usha Rani[1] and Ashwini Kumar[4]

1 Department of Microbiology, Biotechnology and Food Technology, Bangalore University, Bengaluru, India

2 Department of Biotechnology, Manipal Institute of Technology (MAHE-MIT), Manipal, India

3 Nitte University, Center for Science Education and Research (NUCSER), Nitte, (DU), Deralakatte, Mangaluru, India

4 Genei Laboratory Pvt Ltd, Bengaluru, India

*Address all correspondence to: roopa_arekal@yahoo.com

References

[1] Fox PF. Cheese: An overview. In: Fox PF, editor. Cheese: Chemistry, Physics and Microbiology. Second ed. Vol. 1. London: Chapman and Hall; 1993. pp. 1-36

[2] Steinkraus KH. Nutritional significance of fermented foods. Food Research International. 1994;**21**:259-261

[3] Caplice E, Fitzgerald GF. Food fermentations: Role of microorganisms in food production and preservation. International Journal of Food Microbiology. 1999;**50**:131-149

[4] Champagne CP, Mollgaard H. Handbook of fermented functional foods. In: Farnworth ER, editor. Production of Probiotic Cultures and their Addition in Fermented Foods. 2nd ed. Boca Raton, FL: Taylor & Francis Group; 2008. pp. 513-532

[5] Perricone M, Bevilacqua A, Corbo MR, Sinigaglia M. Technological characterization and probiotic traits of yeasts isolated from Altamura sourdough to select promising microorganisms as functional starter cultures for cereal-based products. Food Microbiology. 2014;**38**:26-35. DOI: 10.1016/j.fm.2013.08.006

[6] Di Cagno R, Coda R, De Angelis M, Gobbetti M. Exploitation of vegetables and fruits through lactic acid fermentation. Food Microbiology. 2013;**33**:1-10. DOI: 10.1016/j.fm.2012.09.003

[7] Holko I, Hrabe J, Salakova A, Rada V. The substitution of a traditional starter culture in mutton fermented sausages by *lactobacillus acidophilus* and *Bifidobacterium animalis*. Meat Science. 2013;**94**:275-279. DOI: 10.1016/j.meatsci.2013.03.005

[8] Lavermicocca P, Valerio F, Lonigro SL, De Angelis M, Morelli L, Callegari ML, et al. Study of adhesion and survival of lactobacilli and bifidobacteria on table olives with the aim of formulating a new probiotic food. Applied and Environmental Microbiology. 2005;**71**:4233-4240. DOI: 10.1128/AEM.71.8.4233-4240.2005

[9] Randazzo CL, Todaro A, Pino A, Pitino I, Corona O, Caggia C. Microbiota and metabolome during controlled and spontaneous fermentation of Nocellara Etnea table olives. Food Microbiology. 2013;**65**:136-148. DOI: 10.1016/j.fm.2017.01.022

[10] Roopashri AN, Varadaraj MC. Hydrolysis of flatulence causing oligosaccharides by α-d-galactosidase of a probiotic *lactobacillus plantarum* MTCC 5422 in selected legume flours and elaboration of probiotic attributes in soy-based fermented product. European Food Research and Technology. 2014a;**239**:99-115. DOI: 10.1007/s00217-014-2207-y

[11] Roopashri AN, Varadaraj MC. Soy whey based medium for optimized phytase activity in *Saccharomyces cerevisiae* MTCC 5421 and α-D-galactosidase and antibacterial activities in *lactobacillus plantarum* MTCC 5422 by response surface methodology. Journal of Food Science and Technology. 2014b;**51**(3):519-526. DOI: 10.1007/s13197-011-0527-5

[12] Carr FJ, Chill D, Maida N. The lactic acid bacteria: A literature survey. Critical Reviews in Microbiology. 2002;**28**:281-370. DOI: 10.1080/1040-840291046759

[13] Gerritsen J, Smidt H, Rijkers GT, de Vos WM. Intestinal microbiota in

human health and disease: The impact of probiotics. Genes and Nutrition. 2011;**6**:209-240. DOI: 10.1007/s12263-011-0229-7

[14] O'Sullivan GC, Kelly P, O'Halloran S, Collins C, Collins JK, Dunne C, et al. Probiotics: An emerging therapy. Current Pharmaceutical Design. 2005;**11**:3-10. DOI: 10.2174/13816 12053382368

[15] FAO/WHO. Probiotics in Food: Health and Nutritional Properties and Guidelines for Evaluation; Italy: FAO Food and Nutrition Paper; 2006. p. 85

[16] Monteagudo-Mera A, Rodriguez-Aparicio L, Rua J, Martinez-Blanco H, Navasa N, Garcia-Armesto MR, et al. In vitro evaluation of physiological probiotic properties of different lactic acid bacteria strains of dairy and human origin. Journal of Functional Foods. 2012;**4**:531-541. DOI: 10.1016/j.jff.2012.02.014

[17] Zheng J, Wittouck S, Salvetti E, Franz CMAP, Harris HMB, Mattarelli P, et al. A taxonomic note on the genus *lactobacillus*: Description of 23 novel genera, emended description of the genus *lactobacillus Beijerinck* 1901, and union of *Lactobacillaceae* and *Leuconostocaceae*. International Journal of Systematic and Evolutionary Microbiology. 2020;**70**(4):2782-2858. DOI: 10.1099/ijsem.0.004107

[18] Steinkraus KH. Lactic acid fermentation in the production of foods from vegetables, cereals and legumes. Antonie Van Leeuwenhoek. 1983;**49**:337-348

[19] Jeyaram K, Singh A, Romi W, Devi AR, Singh WM, Dayanithi H, et al. Traditional fermented foods of Manipur. Indian Journal of Traditional Knowledge. 2009;**8**(1):115-121

[20] Sekar S, Mariappan S. Usage of traditional fermented products by Indian rural folks and IPR. Indian Journal of Traditional Knowledge. 2007;**6**(1):111-120

[21] Sandholm TM, Lahteenmaki L, Saarela M. Eu perspectives on food, gastrointestinal tract functionality and human health. In: Goktepe I, Juneja VK, Ahmedna M, editors. Probiotics in Food Safety and Human Health. Boca Raton, FL: CRC press; 2006. pp. 310-332

[22] Swinbanks D, O'Brien J. Japan explores the boundary between food and medicine. Nature. 1993;**364**:180

[23] Gibson GR, Roberfroid MB. Dietary modulation of the human colonic microbiota: Introducing the concept of prebiotics. J Nutr. 1995;**125**(6):1401-1412

[24] Kearney N, Stanton C Desmond C, Coakley M, Collins JK, Fitzgerald G, et al. Challenges associated with the development of probiotic-containing functional foods. In: Farnworth ER, editor. Handbook of Fermented Functional Foods. 2nd ed. Boca Raton: CRC press; 2008. pp. 25-69

[25] Dunne C, O'Mahony L, Murphy L, Thornton G, Morrissey D, O'Halloran S, et al. *In vitro* selection criteria for probiotic bacteria of human origin: Correlation with *in vivo* findings. American Journal of Clinical Nutrition. 2001;**73**:386S-392S

[26] Lankaputhra WEV, Shah P. Antimutagenic properties of probiotic bacteria and organic acids. Mutation Research. 1998;**397**:169-182

[27] Farnworth ER. The history of fermented foods. In: Farnworth ER, editor. Handbook of Fermented Functional Foods. 2nd ed. Boca Raton: CRC press; 2008. pp. 1-24

[28] Gmeiner M, Kneifel W, Kulbe KD, Wouters R, De Boever P, Nollet L,

et al. Influence of a synbiotic mixture consisting of *lactobacillus acidophilus* 74-2 and a fructooligosaccharide preparation on the microbial ecology sustained in a simulation of the human intestinal microbial ecosystem (SHIME reactor). Applied Microbiology and Biotechnology. 2000;**53**:219-223

[29] Goktepe I. Probiotics as biopreservatives for enhancing food safety. In: Goktepe I, Juneja VK, Ahmedna M, editors. Probiotics in Food Safety and Human Health. 2nd ed. Boca Raton: CRC Press; 2006. pp. 285-307

[30] Cummings JH. Short chain fatty acids in the human colon. Gut. 1981;**22**:763-779

[31] Cummings JH. Production and metabolism of short chain fatty acids in humans. In: Roche AF, editor. Short Chain Fatty Acids: Metabolism and Clinical Importance, Report of the Tenth ROSS Conference on Medical Research. Columbus, OH: Ross Laboratories; 1991. pp. 11-17

[32] Ouwehand AC, Vesterlund, S. Antimicrobial components from lactic acid bacteria. In: Salminen, S., von Wright, A. and Ouwehand, A. (Eds.), Lactic Acid Bacteria: Microbiological and Functional Aspects. New York: Marcel Dekker Inc.; 2004. pp. 375-396

[33] Barrett E, Stanton C, Fitzgerald G, Ross RP. Biotechnology of food cultures for the nutritional enhancement of milk and dairy products. In: Shortt C, O'Brien J, editors. Handbook of Functional Dairy Products. Boca Raton: CRC Press; 2004. pp. 262-280

[34] Casci T, Rastall RA, Gibson GR. Human gut microflora in health and disease: Focus on prebiotics. In: Shetty K, Paliyath G, Pometto AL, Levin RE, editors. Functional Foods and Biotechnology. Boca Raton: CRC Press; 2007. pp. 401-434

[35] Gallaher D, Khil J. The effect of synbiotics on colon carcinogenesis in rats. Journal of Nutrition. 1999;**129**:1483S-1487S

[36] Isolauri E, Sutas Y, Kankaanpaa P, Arvilommi H, Salminen S. Probiotics: Effects on immunity. American Journal of Clinical Nutrition. 2001;**73**:444S-450S

[37] Salminen S, Playne M, Lee YK. Successful probiotic lactobacilli: Human studies on probiotic efficacy. In: Colette Shortt C, O'Brien J, editors. Handbook of Functional Dairy Products. Boca Raton: CRC Press; 2004. pp. 26-46

[38] Schmid K, Schlothauer RC, Friedrich U, Staudt C, Apajalahti J, Hansen EB. Development of probiotic food ingredients. In: Ahmedna M, Goktepe I, Juneja VK, editors. Probiotics in Food Safety and Human Health. Boca Raton: CRC Press; 2005. pp. 35-66

[39] Satish Kumar R, Kanmani P, Yuvaraj N, Paari KA, Pattukumar V, Arul V. Traditional Indian fermented foods: A rich source of lactic acid bacteria. International Journal of Food Sciences and Nutrition. 2013;**64**(4):415-428

[40] Chavan JK, Kadam SS. Nutritional improvement of cereal fermentation. CRC Critical Reviews in Food Science and Nutrition. 1989;**28**:349-400

[41] Mattila-Sandholm T. Development and future of probiotic foods - the European approach. In: Lecture in VTT Symposium 177: Biotechnology in the Food Chain - New Tools and Applications for Future Foods. Helsinki, Finland: VTT Technical Research Centre of Finland; 1998

[42] Hamad AM, Fields ML. Evaluation of protein quality and available lysine of

germinated and ungerminated cereals. Journal of Food Science. 1979;**44**:456-459

[43] Sanni AI, Onilude AA, Ogundoye OR. Effect of bacterial galactosidase treatment on the nutritional status of soy bean seeds and its milk derivative. Nahrung. 1997;**41**:18-21

[44] Gillooly M, Bothwell TH, Charlton RW, Torrance JD, Bezwoda WR, Macphail AP, et al. Factors affecting the adsorption of iron from cereals. British Journal of Nutrition. 1984;**51**:37-46

[45] Haard NF, Odunfa SA, Lee CH, Quintero Ramirez R, Lorence-Quinones A, Wacher-Radarte C. Fermented Cereals: A Global Perspective. United Nations: FAO Agricultural Services Bulletin; 1999. p. 138

[46] Khetarpaul N, Chauhan BM. Effect of fermentation by pure cultures of yeasts and lactobacilli on the available carbohydrate content of pearl millet. Tropical Science. 1990;**31**:131-139

[47] Nout MJR, Motarjemi Y. Assessment of fermentation as a household technology for improving food safety: A joint FAO/ WHO workshop. Food Control. 1997;**8**:221-226

[48] Stewart RB, Getachew A. Investigations on the nature of Injera. Economic Botany. 1962;**16**:127-130

[49] Steinkraus KH. Bio-enrichment: Production of vitamins in fermented foods. In: Wood JB, editor. Microbiology of Fermented Foods. London: Blackie Academic and Professional; 1998. pp. 603-619

[50] Oyewole OB. Lactic fermented foods in Africa and their benefits. Food Control. 1997;**8**:289-297

[51] Campbell-Platt G. Fermented foods: A world perspective. Food Research International. 1994;**27**:253-257

[52] Purushothaman D, Dhanapal N, Rangaswami G. Microbiology and biochemistry of idli fermentation. In: "Symposium on Indigenous Fermented Foods", Bangkok, Thailand. In: Steinkraus KH, editor. (1996)Handbook of Indigenous Fermented Foods. 2nd ed. New York: Marcel Dekker, Inc.; 1977. p. 149

[53] Ramakrishnan CV. The use of fermented foods in India. In "Symposium on Indigenous Fermented Foods", Bangkok, Thailand. In: Steinkraus KH, editor. (1996)Handbook of Indigenous Fermented Foods. 2nd ed. New York: Marcel Dekker, Inc.; 1977. p. 149

[54] Steinkraus KH, Ayres R, Olek A, Farr D. Biochemistry of *saccharomyces*. In: Steinkraus KH, editor. Handbook of Indigenous Fermented Foods. New York: Marcel Dekker; 1993. pp. 517-519

[55] Tamime AY, Robinson RK. Fermented milks and their future trends, part II: Technological aspects. Journal of Dairy Research. 1988;**55**:281-307

[56] Rati Rao E, Vijayendra SVN, Varadaraj MC. Fermentation biotechnology of traditional foods of the Indian subcontinent. In: Shetty K, Paliyath G, Pometto A, Levin RE, editors. Food Biotechnology. 2nd ed. Boca Raton: CRC Press; 2006. pp. 1759-1794

[57] Sinha PR, Sinha RN. Importance of good quality dahi in food. Indian Dairyman. 2000;**52**:45-47

[58] Laxminarayana H, Nambudripad VKN, Lakshmi N, Anantaramaiah SN, Sreenivasamurthy V. Studies on dahi, II: General survey of the quality of market dahi. Indian Journal of Veterinary Science and Animal Husbandary. 1952;**22**:13-25

[59] Baisya RK, Bose AN. Role of inoculating organisms on the

physico-chemical changes in milk and final curd quality. Indian Journal of Dairy Science. 1975;**28**:179-183

[60] Sharma CK, Jain SC. Effects of starter cultures and incubation (period and temperature) on the acidity of dahi (curd), part II. Journal of Food Science and Technology. 1975;**12**:81-83

[61] Vijayendra SVN. Preparation, Evaluation and Use of Bulk Freeze-Dried Starter Cultures in Dahi and Yoghurt Manufacture [Thesis]. Karnal, India: National Dairy Research Institute; 1994

[62] Balasubramanyam BV, Varadaraj MC. Dahi as a potential source of lactic acid bacteria active against foodborne pathogenic and spoilage bacteria. Journal of Food Science Technology. 1994;**31**:241-243

[63] Balasubramanyam BV, Varadaraj MC. Antibacterial effect of *lactobacillus* spp. on foodborne pathogenic bacteria in an Indian milk-based fermented culinary item. Cultured Dairy Products Journal. 1995a;**30**:22-27

[64] Yadav H, Jain S, Sinha PR. Evaluation of changes during storage of probiotic dahi at 78C. International Journal of Dairy Technology. 2007;**60**(3):205-210

[65] Tamang JP, Dewan S, Olasupo NA, Schillinger V, Holzapful H. Identification and enzymatic profiles of predominant lactic acid bacteria isolated from soft variety *chhurpi*, a traditional cheese typical of the Sikkim Himalayas. Food Biotechnology. 2000;**14**(1&2):99-112

[66] Prashant, Tomar SK, Singh R, Gupta SC, Arora DK, Joshi BK. Phenotypic and genotypic characterization of lactobacilli from Churpi cheese. Dairy Science & Technology. 2009;**89**:531-540

[67] Thapa TB. Diversification in Processing and Marketing of Yak Milk Based Products. Nepal: TAAAS/IYIC/Yak foundation/FAOROAP/ ICIMOD/ILR/ Workshop; 2002. pp. 484-489

[68] Dewan S, Tamang JP. Dominant lactic acid bacteria and their technological properties isolated from the Himalayan ethnic fermented milk products. Antonie Van Leeuwenhoek. 2007a;**92**:343-352

[69] Dewan S, Tamang JP. Microbial and analytical characterization of Chhu-A traditional fermented milk product of the Sikkim Himalayas. Journal of Scientific and Industrial Research. 2007b;**65**:747-752

[70] Gupta RC, Bimlesh M, Joshi VK, Prasad DN. Microbiological, chemical and ultra structural characteristics of mishti doi (sweetened dahi). Journal of Food Science and Technology. 2000;**37**:54-57

[71] Mann B, Joshi VK. Proteose-peptone contents of mishti doi and their correlation with browning index. Journal of Food Science and Technology. 1997;**34**:425-426

[72] Rangappa KS, Achayya KT. Indian Dairy Products. New Delhi: Asia Publishing House; 1974. pp. 119-123

[73] Bhanumurthy JL, Trehan KS. Preservation and bottling of buttermilk. Indian Dairyman. 1970;**22**:275-278

[74] Naresh K, Prasad DN. Preservative action of nisin in lassi under different storage temperatures. Indian Journal of Animal Science. 1996;**66**:525-528

[75] Patidar SK, Prajapathi JB. Standardization and evaluation of Lassi prepared using *lactobacillus acidophilus* and *Streptococcus thermophilus*. Journal

of Food Science and Technology. 1998;**35**:428-431

[76] Patel RS, Abdel-Salam MH. Shrikhand: An Indian analogue of western quarg. Cultured Dairy Products Journal. 1986;**21**:6-7

[77] Reddy NR, Sathe SK, Pierson MD, Salunkhe DK. Idli, an Indian fermented food: A review. Journal of Food Quality. 1981;**5**:89-101

[78] Subramanian BS, Naresh Kumar C, Narasimhan R, Shanmugam AM, Khan MMH. Selection of level and type of LAB starter in the preparation of dietetic shrikhand. Journal of Food Science and Technology. 1997;**34**:340-342

[79] Boghra VR, Mathur ON. Physico-chemical status of major milk constituents and minerals at various stages of Shrikhand preparation. Journal of Food Science and Technology. 2000;**37**:111-115

[80] Prajapati JP, Upadhyay KG, Desai HK. Comparative quality appraisal of heated shrikhand stored at ambient temperature. Australian Journal of Dairy Technology. 1993a;**47**:18-22

[81] Prajapati JP, Upadhyay KG, Desai HK. Quality appraisal of heated shrikhand stored at room temperature. Cultured Dairy Products Journal. 1993b;**28**:16-17

[82] De A, Patel RS. Technology of Shrikhand powder. Cultured Dairy Products Journal. 1990;**25**:21-28

[83] Mahajan BM, Mathur ON, Bhattacharya DC, Srinivasan MR. Production and shelf life of spray dried shrikhand powder. Journal of Food Science and Technology. 1979;**16**:9-10

[84] Campbell-Platt G. Fermented Foods of the World: A Dictionary and Guide. London: Butterworths; 1987

[85] Steinkraus KH. Handbook of Indigenous Fermented Foods. New York: Marcel Dekker; 1996

[86] Wang HL, Hesseltine CW. Use of microbial cultures: Legume and cereal products. Food Technology. 1981;**35**:79-83

[87] Salovaara H. Lactic acid bacteria in cereal based products. In: Salminen, S. and von Wright, A. (Eds.), Lactic Acid Bacteria. New York: Marcel Dekker; 1993. pp. 111-126

[88] Samanta AK, Kolte AP, Senani S, Sridhar M, Jayapal N. Prebiotics in ancient Indian diets. Current Science. 2011;**101**(1):43-46

[89] Swennen K, Courtin CM, Delcour JA. Non-digestible oligosaccharides with prebiotic properties. Critical Reviews in Food Science and Nutrition. 2006;**46**:459-471

[90] Nout MJR, Sarkar PK. Lactic acid food fermentation in tropical climates. Antonie Van Leeuwenhoek. 1999;**76**:395-401

[91] Satish Kumar R, Kanmani P, Yuvaraj N, Paari KA, Pattukumar V, Arul V. *Lactobacillus plantarum* AS1 binds to cultured human intestinal cell line HT-29 and inhibits cell attachment by enterovirulent bacterium *Vibrio parahaemolyticus*. Letters in Applied Microbiology. 2011;**53**:481-487

[92] Satish Kumar R, Kanmani P, Yuvaraj N, Paari KA, Pattukumar V, Thirunavukarasu C, et al. *Lactobacillus plantarum* AS1 isolated from south Indian fermented food Kallappam suppress 1, 2-dimethyl hydrazine (DMH) induced colorectal cancer in male Wistar rats. Applied Biochemistry and Biotechnology. 2012;**166**:620-631

[93] Antony U, George ML, Chandra TS. Inhibition of *salmonella typhimurium*

and *Escherichia coli* by fermented finger millet (*Eleusine coracana*). World Journal of Microbiology and Biotechnology. 1998;**14**:883-886

[94] Mbithi-Mwikya S, Camp JV, Mamiro PRS, Ooghe W, Kolsteren P, Huyghebaert A. Evaluation of the nutritional characteristics of a finger millet based complementary food. Journal of Agricultural and Food Chemistry. 2002;**50**:3030-3036

[95] Aliya S, Geervani P. An assessment of the protein quality and vitamin B content of commonly used fermented products of legumes and millets. Journal of the Science of Food and Agriculture. 1981;**32**:837-842

[96] Rajyalakshmi P, Geervani P. Studies on tribal foods of South India: Effect of processing methods on the vitamin and in-vitro protein digestibility of cereals, millets and legumes. Journal of Food Science and Technology. 1990;**27**:260-263

[97] Ramakrishnan CV. Studies on Indian fermented foods. Baroda Journal of Nutrition. 1979;**6**:1-7

[98] Desikachar HSR, Radhakrishnamurthy R, Ramarao G, Kadkol SB, Srinivasan M, Subrahmanyan V. Studies on idli fermentation, I: Some accompanying changes in the batter. Journal of Scientific and Industrial Research. 1960;**19C**:168-172

[99] Lewis YS, Johar DS, Subrahmanyan V. Studies on process simplification in the preparation of a fermented type of foodstuff idli. Central Food Technological Research Institute Bulletin. 1955;**5**:257-265

[100] Radhakrishnamurthy R, Desikachar HSR, Srinivasan M, Subrahmanyan V. Studies on idli fermentation, II: Relative participation of blackgram flour and rice semolina in the fermentation. Journal of Scientific and Industrial Research. 1961;**20**:342-345

[101] Steinkraus KH, van Veen AG, Thiebeau DB. Studies on idli: An Indian fermented blackgram-rice food. Food Technology. 1967;**27**:110-113

[102] Thyagaraja N, Otani H, Hosono A. Studies on microbiological changes during the fermentation of idli. Lebensmittel-Wissenschaft Technologie. 1992;**25**:77-79

[103] Yajurvedi RP. Microbiology of idli fermentation. Indian Food Packer. 1980;**34**:33-38

[104] Mukherjee SK, Albury MN, Pederson CS, van Veen AG, Steinkraus KH. Role of *Lecuconostoc mesenteroides* in leavening the batter of idli, a fermented food of India. Applied Microbiology. 1965;**13**:227-231

[105] Soni SK. Studies on some Indian Fermented Foods: Microbiological and Biochemical Aspects [Thesis]. Amritsar, India: Guru Nanak Dev University; 1987

[106] Soni SK, Sandhu DK. Indian fermented foods: Microbiological and biochemical aspects. Indian Journal of Microbiology. 1990;**30**:135-157

[107] Thygaraja N, Otani H, Hosono A. Microflora in idli, a traditional fermented cereal-pulse product from India. Lebensmittel-Wissenschaft Technologie. 1991;**24**:504-507

[108] Venkatasubbaiah P, Dwarakanath CT, Sreenivasa Murthy V. Microbiological and physicochemical changes in idli batter during fermentation. Journal of Food Science and Technology. 1984;**22**:59-63

[109] Van Veen AG, Hackler LR, Steinkras KH, Mukherje SK. Nutritive

value of idli, a fermented food of India. Journal of Food Science. 1967;**32**:339-341

[110] Lakshmi I. Studies on Fermented Foods [Thesis]. Baroda: MS University of Baroda; 1978

[111] Varadaraj MC, Vijayendra SVN, Rati ER, Agrawal R, Prasad MS, Nand K. An Improved Process for the Preparation of Inoculum for Fermented Foods. India: Indian Patent Application 352/DEL/01; 2001

[112] Varadaraj MC, Rati ER, Agrawal R, Vijayendra SVN, Prasad MS, Nand K. An improved process for the preparation of shelf stable idli batter. Indian Patent Application. 1999;**19**:24-86

[113] Agrawal R, Rati ER, Vijayendra SVN, Varadaraj MC, Prasad MS, Nand K. Flavour profile of idli batter prepared from defined microbial starter cultures. World Journal of Microbiology and Biotechnology. 2000;**16**:687-690

[114] Jama YH, Varadaraj MC. Antibacterial effect of plantaricin LP84 on foodborne pathogenic bacteria occurring as contaminants during idli batter fermentation. World Journal of Microbiology and Biotechnology. 1999;**15**:27-32

[115] Soni SK, Sandhu DK, Vilkhu KS. Studies on dosa, an indigenous Indian fermented food: Some biochemical changes accompanying fermentation. Food Microbiology. 1985;**2**:175-181

[116] Soni SK, Sandhu DK, Vilkhu KS, Kamra N. Microbiological studies on dosa fermentation. Food Microbiology. 1986;**3**:45-53

[117] Soni SK, Sandhu DK. Nutritional improvement of Indian dosa batters by yeast enrichment and black gram replacement. Journal of Fermentation and Bioengineering. 1989;**68**:52-55

[118] Neeta Joshi S, Godbole H, Kanekar P. Microbial and biochemical changes during dhokla fermentation with special reference to flavour compounds. Journal of Food Science and Technology. 1989;**26**:113-115

[119] Balasubramanyam BV. Studies on the Application of Antagonistic Lactic Acid Bacteria in the Biopreservation of Selected Indigenous Milk and Cereal/ Pulse Based Foods [Thesis]. Mysore, India: University of Mysore; 1996

[120] Varadaraj MC, Balasubramanyam BV, Joseph R. Antibacterial effect of *lactobacillus* spp. on spoilage bacteria in selected Indian fermented foods. In: Proceedings of International Conference on Traditional Foods. Mysore: Central Food Technological Research Institute; 2000. pp. 210-226

[121] Sukumar G, andGhosh, A.R. Pediococcus spp. – A potential probiotic isolated from Khadi (an Indian fermented food) and identified by 16S rDNA sequence analysis. Afr Journal of Food Science. 2010;**4**(9):597-602

[122] Sandhu DK, Soni SK. Microflora associated with Indian Punjabi Warri fermentation. Journal of Food Science and Technology. 1989;**26**:21-25

[123] Dhankher N, Chauhan BM. Technical note: Preparation, acceptability and B vitamin content of Rabadi – A fermented pearl millet food. International Journal of Food Science and Technology. 1987;**22**:173-176

[124] Gupta M, Khetarpaul N, Chauhan BM. Effect of Rabadi fermentation on HCl-extractability of minerals of wheat. Food Chemistry. 1991;**42**:111-117

[125] Khetarpaul N, Chauhan BM. Improvement in HCl-extractability of minerals from pearl millet (*Pennisetum typhoideum*) by fermentation with yeasts and lactobacilli. Journal of the Science of Food and Agriculture. 1989;**49**:117-121

[126] Khetarpaul N, Chauhan BM. Effect of pure and sequential culture fermentation by yeasts and lactobacilli on HCl-extractability minerals from food millet. Food Chemistry. 1991;**39**:347-355

[127] Mahajan S, Chauhan BM. Phytic acid and extractable phosphorus of pearl millet flour as affected by natural lactic acid fermentation. Journal of the Science of Food and Agriculture. 1987;**41**:381-386

[128] Mahajan S, Chauhan BM. A research note: Effect of natural fermentation on the extractability of minerals from pearl millet flour. Journal of Food Science. 1988;**53**:1576-1577

[129] Gupta V, Nagar R. Effect of cooking, fermentation, dehulling and utensils on antioxidants present in pearl millet rabadi – A traditional fermented food. Journal of Food Science and Technology. 2010;**47**:73-76

[130] Modha H, Pal D. Optimization of Rabadi-like fermented milk beverage using pearl millet. Journal of Food Science and Technology. 2011;**48**:190-196

[131] Montet D, Loiseau G, Zakhia N, Mouquet C. Fermented fruits and vegetables. In: Joshi VK, Pandy A, editors. Biotechnology: Food Fermentations. Vol. 2. New Delhi: Educational Publisher and Distributors; 1999. pp. 951-969

[132] Tamang B, Tamang JP. Traditional knowledge of biopreservation of perishable vegetables and bamboo shoots in Northeast India as food resources. Indian Journal of Traditional Knowledge. 2009;**8**(1):89-95

[133] Singh A, Singh RK, Sureja AK. Cultural significance and diversities of ethnic foods of Northeast India. Indian Journal of Traditional Knowledge. 2007;**6**(1):79-94

[134] Reddy KBPK, Raghavendra P, Kumar BG, Misra MC, Prapulla SG. Screening of probiotic properties of lactic acid bacteria isolated from Kanjika, an ayruvedic lactic acid fermented product: An in-vitro evaluation. The Journal of General and Applied Microbiology. 2007;**53**:207-213

[135] Sura K, Garg S, Garg FC. Microbiological and biochemical changes during fermentation of kanji. Journal of Food Science and Technology. 2001;**38**:165-167

[136] Kingston JJ, Radhika M, Roshini PT, Raksha MA, Murali HS, Batra HV. Molecular characterization of lactic acid bacteria recovered from natural fermentation of beet root and carrot kanji. Indian Journal of Microbiology. 2010;**50**:292-298

Chapter 2

Lactobacillus Use for Plant Fermentation: New Ways for Plant-Based Product Valorization

Morgan Le Rouzic, Pauline Bruniaux, Cyril Raveschot, François Krier, Vincent Phalip, Rozenn Ravallec, Benoit Cudennec and François Coutte

Abstract

Today, plant production is increasing, but most industrial processes generate a lot of waste and by-products for which, in the current context, it is a priority to recycle or valorize them. One of the cheapest valorization routes is fermentation, in particular lactic fermentation by *Lactobacillus* species, which produces lactic acid and other molecules of industrial interest such as bioactive compounds such as anthocyanin, organic acid, peptides, or phenol, which are widely found in the plant matrix, mainly in cereals, grass, fruits, and vegetables. Bioactive compounds may exert beneficial health effects, such as antioxidant, anti-inflammatory, antimicrobial, or prebiotic activities. In addition, lactic acid fermentation can improve existing products and lead to new applications in food, livestock feeding and biotechnology, such as the production of lactic acid, protein, or silage. This chapter reviews the use of *Lactobacillus* strains in the fermentation process of many plant bioresources or by-products through their different bioactivities, active molecules, and applications.

Keywords: *lactobacillus* genera, lactic acid fermentation, by-product valorization, bioactivities, health benefits

1. Introduction

The world's population of 7.6 billion people is still growing and is expected to reach 8.3 billion by 2025 and almost 10 billion by 2050 [1]. Concomitantly, the Earth's resources are depleting. According to the different scenarios, global food demand is expected to increase by 40–68% by 2050 [2]. Among food resources, plants are of particular interests, as the global production of plant-based products is constantly increasing while producing significant waste. In this context, recycling or revalorizing these by-products is a priority [1].

The main objectives of using plant by-products are to revalorize wastes, reduce pollution, and limit resource depletion. Fermentation is one of the least polluting methods. Plant by-products fermentation contributes to sustainable development; in fact, this type of valorization is part of some objectives of the United Nations 2030 Agenda, notably the third objective: good health and well-being and the twelfth objective: responsible consumption and production. The consumption of fermented plant by-products therefore allows responsible consumption. The fermentation of plant by-product leads to bioactivities related to human health such as antioxidant, anti-inflammatory, or antimicrobial activities that contribute to good health and well-being [3]. Plant-based foods are sources of many bioactive compounds such as fibers, vitamins, minerals, or phenolic compounds. These nutrients are necessary for the survival and growth of organisms [4]. In many countries, the health benefits of certain plant and their traditional use have been recognized for decades [5]. Since industries have been exploiting plant-based foods, many agro-industrial by-products that still contain valuable compounds have been generated. Many companies are now seeking to recycle waste from their fruit and vegetable activities in order to address environmental and economic issues. For example, cereal waste reached about 40,000–45,000 tons per year in Europe [6]. The by-products are mainly used for livestock feed or methanisation but have great potential to generate food or dietary supplements for human use [6, 7]. Another example concerns the waste from the citrus industry, which amounts to 50 million tons per year and is the most important waste from fruits exploitation. The management of by-products represents a real food waste problem and raises major issues [8]. Therefore, in recent years, there has been a growing interest in the valorization of plant by-products.

In China, for 9000 years, humans have empirically exploited the fermentation process for numerous applications [9]. Studied since the nineteenth century, lactic acid fermentation has been an essential process for food processing and preservation for many millennia [10]. Humans took advantage of it for their food, notably by developing bread, beer, wine, cheese, or vinegar. Subsequently, fermentation with lactic acid bacteria has been largely studied to improve the nutritional and functional properties of plants. Due to their richness in nutrients, water, and natural ferments, plants such as fruits and vegetables represent an optimal substrate for *Lactobacillus* [11]. Lactic acid bacteria constitute a diverse group of Gram-positive, catalase-negative bacteria producing lactic acid as the main end product. Many food products fermented by lactic acid bacteria are obtained with organisms belonging to the genus *Lactobacillus* [12]. With more than 200 species of *Lactobacillus* bacteria [11], this genus is certainly the main and most diverse group of lactic acid bacteria. A study published in 2020 re-evaluated the genetic relatedness and phylogeny of *Lactobacillus* species. Based on a polyphasic approach such as whole-genome comparison, core genome phylogeny, physiological criteria, and ecology of the organisms, the genus *Lactobacillus* was reclassified into 25 genera (2 preexisting genus and 23 new genera). This work showed the great and extensive diversity of the *Lactobacillaceae* family [13]. *Lactobacillus* species are commonly used in fermented food. Depending on the species, their enzymatic activities including amylase, lactate dehydrogenase, peptidase, proteinase, α- and β-glucosidases, decarboxylase, lactate dehydrogenase, peptidase, phenolic acid decarboxylase, phenol reductase, proteinase or tannase are very useful in food fermentation [14]. These enzymes can degrade the plant cell wall matrix, resulting in the release of many bioactive compounds, which may or may not be modified structurally by the action of other enzymes in the bacteria.

Today, several ecological and economic issues are at the heart of lactic fermentation research. The optimization of yield, cost, and energy consumption and the valorization of plant-derived products represent challenges for the industry [15]. To meet this demand, the use of new substrates and the genetic engineering of fermentation strains are being studied as potential solutions [11]. Moreover, it is now known that lactic fermentation increases the content of bioactive compounds. Indeed, this fermentation process is well known to strengthen the immune and antioxidant (AO) effect of medicinal plants by increasing the bioavailability of active compounds, but also through the production (or the bioconversion) of plant metabolites into new bioactive molecules [16]. To increase the bioactivities and organoleptic characteristics of fermented products, *Lactobacillus* converts metabolizable molecules with their enzymes, in particular *L. plantarum*, which is one of the most used *Lactobacillus* as a fermentation starter. This degradation increases the bioavailability of molecules and improves their absorption [17]. A fermentation starter is usually a consortium of bacteria that helps the fermentation process to start. Today, the use of starter cultures in food fermentation is one of the necessary ingredients for good production. In addition, LAB used as starter in the food industry provide safe product with good nutritional and organoleptic qualities. LAB are used as starter for many products, including fruit, vegetables, and cereal [18]. As illustrated in **Figure 1**, the production of biomolecules by lactic fermentation of plant by-products can induce other bioactivities. This chapter refers to antioxidant (AO), anti-inflammatory (AI), antimicrobial, prebiotic activities, and others. These can be used in human food and beverage, livestock feeding, or biotechnology mainly to produce lactic acid. Those activities and applications will be detailed in this chapter.

2. Bioactivities resulting from the fermentation of plant products or by-products by *Lactobacillus* genera

2.1 Antioxidant activity

Many *Lactobacillus* enzymes can generate compounds with strong AO activity from plant by-products. For example, β-galactosidase releases isoflavone and oleuropein aglycone while tannases generate propylgallate [16].

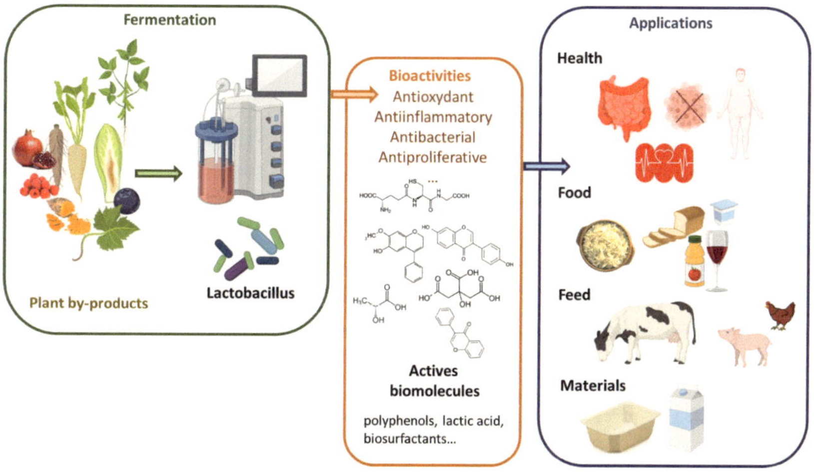

Figure 1.
Summary of the biomolecules, bioactivities generated by the fermentation by Lactobacillus *strains of plant products or by-products, and their application domains.*

Glycosylated polyphenols such as tannins, lignans, isoflavones, flavonols, and anthocyanins are widespread in plant products. Absorption in the intestine depends mainly on their degree of glycosylation. Some strains of *Lactobacillus,* such as *L. plantarum*, possess glycosidases that are crucial for the absorption of glycosylated polyphenols and consequently for the resulting AO activity [19]. In most cases, the AO activity is studied with classical biochemical antioxidant assays such as 2,2-diphenyl-1-picrylhydrazyl (DPPH), 2,2'-azino-bis (3-ethylbenzothiazoline-6-sulfonic acid) (ABTS), hydrolxyl or alkyl radical scavenging activities, the ferric-reducing antioxidant power (FRAP), superoxide dismutase (SOD) -like activity, β-carotene bleaching, and oxygen radical absorbance capacity (ORAC). In addition to *in vitro* biochemical tests, other studies have investigated the antioxidant capacity of fermented products with *in vitro* cell-based assays. Reference [16] demonstrated that the fermentation of *L. plantarum* increased the AO properties of a kiwi extract. They correlated this result with increased amounts of protocatechic and chlorogenic acid in the fermented products, which were less represented in the starting extract [20]. Gallic acid production was also observed with the fermentation of red chicory leaves by *L. plantarum* et *L. hilgardi* thanks to tannases [21]. In addition, co-fermentation by *L. gassieri* and *Bifidobacterium animalis* resulted in the release of caffeic acid and conjugated chlorogenic acid after fermentation of sunflower seeds through the action of cinnamoyl esterase. Tannins are also the product of biomass fermentation by *Lactobacillus.* Tannases hydrolyze the ester bond, and gallate decarboxylase converts gallic acid to pyrogallol; thus, *Lactobacillus* generates gallic acid, glucose, and pyrogallol [22].

Several studies have illustrated the fermentation of plants such as Indian chilli pepper, grape pomace, dandelion beverage, and cereal-based plant beverages by *Lactobacillus spp*., resulting in polyphenol compounds (caffeic acid, succinate, pyruvate, pyroglutamate) with AO capacity [23–25]. In [26], they evidenced that rice bran and wheat bran fermented with *L. plantarum* possessed AO capacity through their hydroxyl and oxygen radical scavenging activities. Furthermore, the purified fractions exerted reactive oxygen species (ROS) scavenging activity in HUVEC cells and decreased the senescence of the cultured cells, also conferring an antiaging activity to the fermented fractions. These activities were attributed to the acids and ketones [26]. Co-cultivation of *L. johnsonii* and *Bacillus coagulans* was undertaken in [27] to produce a soybean meal with improved AO properties. Interestingly, the co-cultivation resulted in a significant increase in total phenolic content [27]. Fruits are also an excellent matrix for fermentation due to their high content of dietary fiber, sugars, vitamins, minerals, and phenols. Furthermore, lactic fermentation preserves and improves food safety, nutritional value and preserves the organoleptic quality. In particular, when plants are fermented by *Lactobacillus* endophyte, it preserves of color, firmness, AO activity, growth of fermentation starters and inhibits pathogens in media. Many studies have been conducted on the lactic fermentation of polyphenol-rich berries and red fruits. *L. casei* has been studied for the fermentation of blueberry pulp [28]. In another example studied in [29], mulberry juice fermented in coculture by three different strains (*L. plantarum, L. acidophilus*, and *L. paracasei*) showed a higher AO capacity [29].

In reference [30], they investigated the fermentation of cherry silverberry fruits (*Elaeagnus multiflora Thunb*.) fermented with pure cultures of *L. plantarum* KCTC 33131 and *L. casei* KCTC 13086 alone or in mixed culture. In reference [31], they studied the fermentation by *L. plantarum* FNC 0027 of Jamaican cherry (*Muntingia calabura Linn.*), which induces the production of phenolic compounds and the inhibition of diabetic-related enzymes (α-glucosidase, α-amylase, and amyloglucosidase). They demonstrated the production of gallic acid, 5,7 dihydroxyflavone, and dihydrokaempferol [31].

The valorization of argan press cake was also carried out by lactic acid fermentation using a specifically isolated strain of *L. plantarum* Argan-L1. Argan press cake is a waste of oil production containing polyphenols and saponins. The authors demonstrated that sucrose from argan press cake was easily converted to lactic acid during the fermentation process. Furthermore, the fermented extract presented an increased AO capacity, but the total phenolic compound was slightly decreased [32].

L. plantarum KCCM 11613P isolated from Kimchi allowed the production of ginsenosides after fermentation of Korean red ginseng (*Panax ginseng*) [33]. In reference [34], it was shown that fermented soymilk products exhibited improved AO capacity associated with increased isoflavone aglycone content. In addition, fermented extracts inhibited the DNA oxidation induced by the Fenton reagent [34]. All these studies show the interest in using *Lactobacillus* to increase the antioxidant properties of fermented products. Moreover, this antioxidant activity is often associated with the anti-inflammatory activity of certain extracts. Fermentation of other plant matrices can induce antioxidant activity of the products, as shown in **Table 1**.

2.2 Anti-inflammatory activity

Vegetables, fruits, and plants (tomato, cucumber, pear, apple, mandarin, parsley, carrot, celery, onion, burdock, kale, spinach, aloe vera, civet, grape, jujube, cabbage, and perilla) fermented by *L. plantarum* offer interesting AI molecules [51]. These molecules include organic acids (OAs) such as lactic acid, 3-phennyl-lactate, indole-3-lactate, β-hydroxybutyrate, gamma-aminobutyric acid (GABA), and glycerol. When investigating the AI (and AO) capacity of these compounds, the parameters studied werethe levels of nitric oxide (NO), IL-6 (interleukins) and tumor-necrosis factor-alpha (TNF-alpha), and the DPPH test on RAW cells [52]. Another study showed the AI properties of a fermented plant extract (*Artemisia capillaris*) in RAW 264.7 cells, which stimulated NO and IL-10 secretion without cytotoxic effects [53]. Thus, the fermentation of *Aronia melanocarpa* extract by *L. plantarum* was investigated to produce GABA, polyphenol, and flavonoid compounds. The fermented extract was shown to exert AI effects inhibiting the production of proinflammatory cytokines in RAW 264.7 cells and modulating the immune response in mice [54]. Furthermore, several molecules derived from the fermentation of red fruit juices have been studied for their AI effects. For example, anthocyanins from these products are thought to produce the TNF-alpha and proinflammatory cytokines [23].

Fermented Asian products were highly investigated for their AI properties. For example, a specific strain of *L. plantarum* is involved in the fermentation of the traditional korean fermented vegetable food, the kimchi. It has been shown to secrete exopolysaccharides able to protect against rotavirus-induced diarrhea [55]. Turmeric, another plant originating from Asia, has been also extensively studied for its AI properties and particularly after fermentation. The development of turmeric extracts with potential health applications, particularly for inflammation, is increased.

The production of curcuminoid molecules, such as curcumin, has been enhanced by fermentation of turmeric (*Curcuma longa*) by *L. johnsonii*. The turmeric extracts showed AI and antiallergic effects in atopic dermatitis mice and induced a decrease in serum immunoglobulin E and proinflammatory cytokines in lipopolysaccharide-induced inflammation (LPS) [56]. Supplementation of turmeric extract fermented by *L. rhamnosus* (GG-ATCC 53103) and *Bifidobacterium animalis* (BB12) strains maintained bacterial growth of the gut microbiota in case of inflammation.

By-product used	*Lactobacillus spp.*	**FP**	**Product generated**	**Bio activity**	**Remark**	**Reference**
Apple juice	*L. plantarum*	LF	PC	AO		[35]
Apple juice	*Saccharomyces cerevisiae,* then *L. plantarum*	LF	PC, OA	AO		[36]
Margosa (*Momordica charantia L.*)	*L. plantarum* NCU116	LF	SCFA LA PC	AO	Juice's sterilization exerted adverse effects	[37]
Porcelain plant (*Graptopetalum paraguayense E. Walther*)	*L. plantarum* BCRC 10357	LF	PC	AO	Assayed during the maturity of the leaves	[38]
Milled wheat	*L. plantarum+ Streptococcus thermophilus*	LF Co	PC	AO, AM, PB	Anti-burning properties	[39]
Apple by-products	*L. plantarum*	LF	PC	AO, barrier integrity	Caco-2	[40]
Mango	*L. plantarum + Saccharomyces cerevisiae*	SB Co	Mango slurry, PC	AO		[41]
Liquorice root	*L. plantarum*	SBF	PC	AO		[42]
Jussara pulp (*Euterpe edulis*)	*Lactobacillus and Bifidobacterium*	LF Co	OA: protocatechic acid	AO	Conversion of anthocyanins	[43]
Acerola	*L. acidophilus +Bifidobacterium longum*	BF	beverage	PB	↗resistance of PB to gastrointestinal digestion	[44]
Cauliflower & white beans mix	L. plantarum 299	SBF	Riboflavin, Folate, Vitamin B12, AA		Nutritional value	[45]
Wheat germ	L. plantarum + L. rossiae	LF	Bread rich in PC, phytases	AO	↘ anti-nutritional factor, ↗ protein digestibility	[46]
Date juice	*L. casei subsp. rhamnosus*	LF	LA		Nitrogen source optimization	[47]
Date juice	*L. sp.* KCP01	LF	LA		Medium optimization	[48]

By-product used	*Lactobacillus spp.*	FP	Product generated	Bio activity	Remark	Reference
Solid carob	*L. rhamnosus*	BF	LA	Many diseases	Immobilization in alginate beads	[49]
Banana, papaya, pineapple, orange	*L. plantarum*	BF	LA		Best LA's production for banana et pineapple	[50]

AI: Anti-inflammatory; AM: antimicrobial; AO: AO; BF: batch fermentation; Co: coculture; EPS: exopolysaccharides; FP: fermentation process; IL: interleukins; IM: immune-modulatory; LA: lactic acid; LF: liquid fermentation; NO: nitric oxide; OA: organic acids; PB: prebiotic effect; PC: phenolic compounds; SBF: solid batch fermentation; SCFA: short-chain fatty acid; and SF: solid fermentation.

Table 1.
Other studies that complement the in vitro examples cited in this chapter.

It also reduced the inflammatory state by limiting the production of proinflammatory cytokines IL-8 [57]. Another study showed that the fermentation of turmeric by *L. fermentum* has increased the curcumin yield by 9.76%. The AI activity was demonstrated in RAW 264.7 cells by modifying the nitrite level, the expression of TNF-alpha and TLR-4, and the activation of the JNK pathway. These phenolic compounds also showed a protective effect against the activation of TLR-4 receptor cascade, TNF-alpha, and nitric oxide production. In addition, the extract limited the proinflammatory response and low-grade oxidative stress induced by LPS [58].

2.3 Antimicrobial activity

The molecules produced during the fermentation of plant biomasses by *Lactobacillus* can also exhibit antimicrobial activities. The production of antimicrobial molecules by *Lactobacillus* has already been described, including lactobrevin and lactobacillin [59]. For example, in [60], an interesting concept of valorization of okara by solid-state fermentation was presented with a coculture of the yeast *Yarrowia lipolytica* and *Lactobacullis casei*. Okara is an oleaginous by-product of plant milk production. The authors used fermentation to generate molecules with antimicrobial activity (up to 33% reduction of *Bacillus subtilis* development and a modest effect on *Aspergillus niger* one) [60].

In reference [61], a metabolic study on *Allium tuberosum* to produce a food additive with antimicrobial activity against poultry pathogens was conducted. Endophytic *Lactobacillus* have been isolated from Chinese chives. Among those *Lactobacillus* strains, *L. plantarum* can produce flavonols with antimicrobial activity [61]. In [62], fermentation of quinoa by the strain *L. plantarum* CDL 778 leads to a higher production of antifungal compounds. It was also observed that during the fermentation of sweet lemon juice (*Citrus limetta*), the antimicrobial activity against *Escherichia coli* and *Salmonella Typhimurium* was increased. These activities were correlated with the increase in lactic acid content and the decrease of citric acid, total phenolic compounds, and sugar content [63]. Moreover, fermentation of the red sorghum cereal allows the conversion of flavanones into eriodyctiol and naringenin, which have shown an interesting antimicrobial activity [22].

2.4 Prebiotic activity

Several studies have shown that fermented fruits and vegetables have prebiotic effects. The compounds produced by the fermentation of plants induce a modification of the intestinal microbiota. These fermented extracts offer great prospects. Studies highlighted their health potential for humans but also animals. Indeed, two fermented extracts obtained from algae and chicory, plantain, alfalfa, and broad leaf dock presented prebiotic and AO effects. This study was conducted on weaned lambs, and the results showed improved resistance to infection and survival for both extracts. Similar studies have shown the same effects for thyme and rosemary [64]. In reference [65], the prebiotic potential was determined, and the AI effect of chicory root and pulp compared with inulin, as a positive control, on the intestinal barrier on IPEC-J2 cells. These tests were performed with five fermented by-products (chicory roots, chicory and citrus pulp, rye bran, and soybean bark) by different *Lactobacillus spp.* An increase of *Lactobacillus* spp. was observed for all substrates except for chicory roots. The latter was very fermentable and produced a butyrate ratio similar to that of inulin, while chicory pulp had a higher ratio than inulin. For acetate, chicory and citrus pulp and soybean bark had a higher ratio than inulin. These short-chain fatty acids (SCFAs) derived from dietary fiber fermentation contribute to maintain intestinal health. Rye bran caused a significant stimulation of the growth of *Bifidobacterium spp.* Rye bran and soybean bark have a positive effect on the gut microbiota. Fermented chicory roots and pulp promote the upregulation of tiny junction genes and maintain the integrity of the gut barrier. Finally, fermented chicory pulp inhibits proinflammatory cytokines such as TNF-alpha and triggers the metabolic pathway that inhibits inflammatory cytokine production [65].

2.5 Other bioactivities related to medicine

Many bioactivities could result from the lactic fermentation of plant by-products. In reference [66], they associated the AO activity with potential hypoglycemic effects of *Diospyros lotus* fruit fermented by *L. plantarum* and *Microbacterium flavum*. They observed an inhibition of the α-glucosidase activity *in vitro*. In addition, the authors showed that catechinic, tannic, and ellagic acid levels were enhanced during fermentation [66]. Similarly, several studies were interested in the capacity of *Lactobacillus* fermented products to exert a positive effect in the prevention of obesity and associated metabolic diseases. In [67], cabbage-apple juice fermented by *L. plantarum* exerted anti-obesity and hypolipidemic effects *in vivo* in high-fat diet–fed rats was highlighted [67].

Moreover, soy products fermented with *Lactobacillus spp.* have interesting biomolecular contents and present antitumoral effects. Indeed, these fermented soybean extracts could inhibit, *in vitro,* the growth of several cancerous cell models: fibrosarcoma and adenocarcinoma of the breast. It also reduces the risk of breast cancer, significantly influencing survival, apoptosis, and tumor inhibition rates in mice. Clinical studies were also conducted to investigate the effects of fermented soybean extract on chemotherapy-induced immunosuppression. The results showed that the populations of immune cells with activity against tumor cells, the natural killer cells, are significantly increased [23]. Using cell-based experiments, other work has investigated putative health effects associated with AO activity. Indeed, the authors showed promising antiproliferative and apoptotic effects of the extracts on the HeLa cancer cell line. In another study, the authors showed that blueberries fermented by *L. plantarum* exhibited anticancer activities. Their results suggest that polyphenols, in high concentrations in blueberries, were metabolized during fermentation into active phenols such as catechol [68].

3. Applications

3.1 Food products

Product of the lactic fermentation, bread has been, for a longtime, an important foodstuff of the diet of many cultures. The bread fermentation process has often been optimized and revisited to better meet consumer needs or to address economic and social issues. The fermentation of wheat leaven by *L. plantarum* allows the conversion of ferulic acid into vinyl guaiacol, ethyl guaiacol, and dihydro ferulic acid. This conversion improves the quality of the final bread product [22].

Corn flour is another example of a bread raw material, and its application in bakery illustrates the potential of lactic fermentation. In addition to the different ingredients of wheat bread, maize meal improves the nutritional profile after being fermented with *L. plantarum* T6B10 and *Weissella onfuse* BAN8. Indeed, an increase in amino acid (AA) and protein content, AO activity, and inhibition of lipases and phytic acid were observed. This leads to an increase in dietary fiber, digestibility, and improves the texture, taste, and nutritional value of bread [69]. The same outcome was observed with the fermentation of brans from hullless barley, emmer and pigmented wheat varieties with the same *Lactobacillus* under the same conditions [70]. Another study highlighted the replacement of wheat flour substitute for breadmaking, a sourdough obtained from fermented djulis (*Chenopodium formosanum*) by *L. casei*. The bread produced contained higher levels of total phenolic and flavonoid compounds and increased hardness and chewiness compared with conventional bread. The addition of djulis sourdough also extended the shelf life by approximately 2 days [71].

A process to valorize semolina pasta with hemp flour, chickpea grains, and milling by-products by fermenting them with *L. plantarum* and *L. rossiae* has been proposed [72]. However, it is necessary to note that enzymatic pretreatment of the substrates must be carried out beforehand. This could affect the economic viability of the process. At a laboratory scale, they obtained extensive protein degradation and consequently digestibility, a 50% reduction in tannin concentration and also in phytic acid concentration [72]. *L. plantarum,* which has high proteolytic activities, was used for the fermentation of quinoa instead of wheat. Quinoa is an interesting cereal for celiac patients because it is gluten-free. The study revealed that quinoa is more easily fermented by lactic acid bacteria than wheat. These high proteolytic activities of the strain were evidenced by the increase of the total peptides and free AA contents from quinoa slurries compared with wheat slurries [62]. In reference [70], the potential use of oat extract from cereal processing with high protein content as an alternative to yoghurt was questioned. Fermentation of this by-product with *L. delbrüchii* subsp. *Bulgaricus* and *Streptococcus themophilus* followed by starch gelatinization by heating generated two kinds of gels with interesting rheological and organoleptic properties. Authors placed their studies in the context of plant-based products substituting dairy ones for health and environmental reasons. They discussed the consumer acceptance of these products but claimed that sensory descriptors such as soft, sweet, and smooth are highlighted by the sensory panel [73]. Another example of food products fermentation value is the fermentation of olive by *L. plantarum*. Kachouri et al. have shown that the phenolic content of olive oil increases after fermentation by this strain [74]. Other studies have shown that the fermentation of the common Spanish table olive improves preservation and the taste. Indeed, *L. plantarum* ferments olive brine, leading to a reduction in the oleuropein content of the olives [75–79]. In addition, wastewater from olive production, which is another olive coproduct, has been exploited in [80]. When fermented by *L. plantarum*, the content of phenolic compounds becomes more interesting. The antioxidant activity was tested by DPPH and ABTS assay. This coproduct has a 50% higher antioxidant activity after fermentation by *L. plantarum* [80].

In order to innovate in the food market, research is being carried out into the development of plant-based drinks rich in active compounds and with health benefits for consumers. Functional plant beverages fermented with *Lactobacillus* are being widely studied. Aqueous extracts of plants such as soy, pea, coconut, or rice represent alternatives to nondairy milk. Lactic acid fermentation of these beverages could improve the protein content, solubility, and availability of AA. Some strains of *Lactobacillus* are also responsible for the biosynthesis of vitamins during fermentation (vitamin K, vitamin B). Anti-nutritional compounds such as phytates are hydrolyzed during fermentation by some phytase-producing strains, which improves the digestibility and mineral content of the final product [81]. However, optimizing flavors and nutritional quality remains a challenge today because the latter are often criticized for their low nutritional quality and bland taste caused by their short shelf life. A color change has been observed by Do and Fan in fruit or carrot juices fermented by *Lactobacillus* strains, indicating that carotenoids are modified to cis-carotenoid isomers responsible for color change during fermentation by *Lactobacillus* [82]. Rheological studies have also been performed. Indeed, in [83], the effects of different *Lactobacillus* species on volatile and nonvolatile flavor compounds in juices fermentation were studied. The main objective of this research was to identify the marker metabolites generated by different species of *Lactobacillus* strains, which contribute to the flavor and reveal the roles of various *Lactobacillus* species in the formation of flavor compounds. The main markers were 2.3 butanedione, hexenal, acetic acid, formic acid as volatile compounds and lactic acid, malic acid, citric acid as nonvolatile compounds [83].

In another application for the beverage sector, one of the main ideas is to provide fermented products with prebiotic effects from a different matrix of vegetable juice as raw material. Consumers' demand for non-dairy prebiotic foods is constantly increasing due to drawbacks related to dairy foods such as allergy, lactose intolerance, as well as lifestyle change or religious beliefs. In this context, reference [71] presents a development of a functional drink based on soy and quinoa (*Chenopodium quinoa Willd*) obtained by fermentation by *Lactobacillus casei* LC-1. This drink presents a prebiotic effect stimulating the gut microbiota and reducing the following bacterial populations: *Clostridium spp, Bacteroides spp, Enterobacteria, and Enterococcus spp* [84]. Cabbage juice and fresh cabbage, fermented by *Lactobacillus*, are also being studied for the development of probiotic products. When mixed with other vegetables (carrots, onion, cucumber), white and red cabbage fermented with *L. plantarum, L. casei , L. acidophilus,* or *L. delbrueckii* shows a good fermentation profile and potential as a functional probiotic drink as demonstrated by Hyunah et al. [85–88]. Dunkley and Hekmat evaluated the sensory properties and worked to assess the growth and viability of *L. rhamnosus* GR-1 in carrot juice, carrot apple juice, carrot orange juice, and carrot beetroot juice over 72 h of fermentation and 30 days of refrigerated storage at 4°C. The conclusion was that carrot, carrot apple, carrot orange, and carrot beetroot juice fermented with *L. rhamnosus* GR-1 proved to be a satisfactory alternative to dairy-based prebiotic products. All juices achieved viable counts above the minimum counts required to be classified as prebiotic. The results of sensory evaluation also indicated a market potential for prebiotic vegetable juice. The development of prebiotic vegetable juice using *L. rhamnosus* GR-1 as a probiotic agent will provide consumers a viable non-dairy alternative that can provide many health benefits [89]. Co- or triculture can be used to enhance activities. Bergamot juice was fermented by three *Lactobacillus* (*L. plantarum* 107 subsp *plantarum* PTCC 1896, *L. plantarum* AF1, *L. plantarum* LP3) in triculture. This combination resulted in a higher AO activity. Bergamot juice fermented could also be used as a functional drink [90].

Other by-products are recycled, especially in the brewery sector. One study aimed to produce a polyphenol-rich beverage from brewers' spent grain. Fermentation by *L. plantarum* ATCC 8014 was realized, followed by tests on phenolic compound content and AO activity. Phenol content and AO have increased during fermentation. The beverage was more concentrated in phenolic compounds than before fermentation, and its bioactive compounds were more stable [91]. More recently, coffee cherry pulp has been used in infusion to obtain an AO drink called cascara. To improve the AO activity of this beverage, it was fermented by endophytic *L. casei* [92]. A turmeric-based functional drink was also obtained by co-fermentation with *Enterococcus faecium, Lactococcus lactis subsp. Lacti, and L. plantarum*. The AO capacity was measured by titration of total phenolic compounds, and the prebiotic effect was also highlighted by *in vitro* and *in vivo* tests.

Kombucha is a sweet infusion of green tea leaves usually fermented with Kombu, a fungus. One study shows that replacing Kombu with *L. casei* and *L. plantarum*, which are derived from kefir, enhances the production of glucuronic acid, leading to greater antimicrobial and antioxidant activities [93]. Another study showed that a mixture of LAB from kefir and kombucha (*L. casei, L. plantarum, L. acidophilus. L. casei,* and *L. plantarum*) increases the glucuronic acid concentration, antimicrobial and antioxidant activities and allows the use of Kombucha as a health drink [94]. Hou et al. demonstrated the link between antimicrobial activities of kombucha with polyphenols and LAB, especially against *Escherichia coli, Salmonella tiphy, Vibrio cholerae,* and *Shigella dysenteriae* [95]. Green tea used in Kombucha may have activity when fermented by *L. plantarum*. Indeed, fermented extract derived from *Camellia sinensis* is able to mitigate ethanol-induced liver damage. *In vitro* and *in vivo* tests on hepatic cells (HepG2,) and murin model exposed to fermented green tea extract show after exposure of ethanol a better viability and an increase of hepatic alcohol dehydrogenase [96].

3.2 Livestock feeding

The products of plant fermentation by *Lactobacillus* strains can be used in many fields ranging from livestock feeding, such as ruminant by decreasing gas production [97]. *Lactobacillus* strains can also be used for silage preparation. Silages are grass or other green fodder that is compacted and stored under airtight conditions, typically in a silo, for use as livestock feeding in the winter. Many studies focus on using *Lactobacillus* strains to improve the quality of the silage. In reference [98], the effect of *L. brevis* and *L. parafarraginis* used as inoculants and the microbial communities of corn stover silage were studied. After 20 days, the two *Lactobacillus* strains were predominant, and a reduction in lactic acid content coupled with an increase in acetic acid and 1.2-propanediol contents was observed. An improvement in the silage quality and reproducibility of the ensiling process were observed [98]. Recently in [99], the effect of *L. plantarum* addition on the nutritive value of dwarf elephant grass (*Pennisetum purpureum* cv Mott) silage was presented. The aim was to examine the effects of different *L. plantarum* addition on the physical quality, pH, and nutritional value (dry matter, organic matter, crude protein, crude fiber). After incubation, a good silage quality (fresh and acidic odor, good texture, and no fungi) and a pH around 4 were observed. *L. plantarum* addition accelerates ensilage fermentation [99]. In [100], an increase in silage quality by adding waste molasses to *L. plantarum* MTD1 was observed. In the same context, the addition of cellulase was studied to evaluate the effects on the chemical composition, bacterial communities, stability of mixed silage made with high-moisture amaranth and rice straw fermented by *L. plantarum*.

Cellulases increased the abundance of *Lactobacillus* bacteria and reduced the abundance of other lactic acid bacteria. It decreased pH, acetic acid content, ammonia nitrogen content and increased lactic acid concentration after 7 days of ensiling [101]. In conclusion, silage treated with both *Lactobacillus* bacteria and cellulase showed the best silage quality. Optimizing the digestibility of feeds and thus increasing their nutritional value are a challenge for the livestock feeding industry. In another study, the fermentation product of a mixture of ginger and turmeric extract by *Lactobacillus* spp. was supplemented to chickens. Biological analyses of AO enzymes and analysis of gut microbiota and lymphoid organs showed a prebiotic effect, an AO effect, and an improvement in resistance to bacterial infections [102].

3.3 Lactic acid production from plant biomass

The use of low-cost by-products is of primary interest as it reduces production costs compared with the use of complex culture media made with pure and refined products. Consequently, many by-products have been tested in the last few years, in association with screening of the best microbial strains, the best fermentation process, and the best conditions to make them work together [103]. Lactic acid is one of the most widely used organic acids for a long time in various industries, such as food, cosmetics, pharmaceutical, and textile industries, and flavor, conservation, AO, and antimicrobial activity [104]. In the last decade, it has also become an essential platform molecule in the biomaterials sector to produce poly-lactic acid (PLA), a bio-based polymer. This new interest has led to an explosion in worldwide demand. One of the characteristics of polylactic acid is its thermal resistance, a critical parameter for manufacturing thermoformed materials (packaging, film, etc.). *Lactobacillus* have been traditionally used for lactic acid production [105, 106]. When using large-scale fermentation bioprocesses, the biomass feedstock must be carefully selected as it accounts for almost half of the biopolymers production costs [105]. To address this production cost issue, scientists and industrials have been focused on lignocellulosic biomass as a fermentation substrate for lactic acid production. Nevertheless, in order to be easily usable, saccharification pretreatments are needed to break down the cellulose into fermentable carbohydrates. Moreover, *Lactobacillus* are classified as either homofermentative or heterofermentative. *L. delbrueckii* is a homofermentative strain commonly used for the production of lactic acid [107]. Homofermentative strains of *Lactobacillus* cannot use pentose carbohydrates from hemicellulose, but heterofermentative ones, such as *L. brevis*, can use these carbohydrates through the phosphoketolase pathway [106].

In reference [105], the fermentation of 11 different carbohydrates from seaweed or plant biomass as a carbon source to produce L-lactic acid with seven different *Lactobacillus* species was investigated. A comparative analysis of the expected yield of lactic acid production revealed that seaweeds provided comparable production rates to lignocellulosic biomasses [105]. In another study, beet molasse was used to produce lactic acid using *L. delbrueckii* IFO 3202 during batch and continuous fermentation, dilution rate of 0.5 h^{-1} was determined to be the best one and allowed to reach a maximum productivity of 11 g L^{-1} h^{-1}. Authors have demonstrated the importance of medium supplementation by yeast extract, as Lactobacilli are tedious microorganisms that require many substrates and substances to grow [108]. Nevertheless, it is estimated that the addition of yeast extract can contribute up to 30% of the cost of producing lactic acid [109]. Zhang & Vadlani studied the production of D-lactic acid by a homofermentative strain, *L. delbrueckii* ATCC 9649, through a sequential hydrolysis and fermentation process (SHF) and a simultaneous saccharification and fermentation process (SSF). In this work, first, the saccharification of pulp and corn stover was done, and then carbohydrates generated from hydrolysis were used by *L. delbrueckii* and converted to D-lactic acid with high purity (99.8 %). The authors highlighted that the SHF process, compared with the SSF process, avoids substrate inhibition and increases the productivity and the yield of D-lactic acid [107]. The same researchers' team has then engineered a strain of *L. plantarum*, introducing gene encoding isomerase and xylulokinase, for the overproduction of D-lactic acid from corn stover and soybean meal extract. In this work, the authors optimized the culture medium through response surface methodology using saccharified corn stover as carbon source and soybean meal extract as a nitrogen source to substitute YE in the medium to produce high purity of D-lactic acid (99%). A maximum productivity of 0.82 g L^{-1} h^{-1} of D-lactic acid was obtained in the optimized medium, 10% higher than with YE as the main nitrogen source [106].

Saccharification and fermentation could be performed at the same time (simultaneous saccharification and fermentation) and have been used for instance by Tu et al. for LA production. With *L. plantarum*, they obtained up to 65.6 g L^{-1} of lactic acid with a cellulose conversion of 69% [110]. Using inulin from chicory, in [111] they obtained a better performance by simultaneous saccharification and fermentation to produce D-lactic acid with *L. bulgaricus*. In their process, they obtained an optically pure molecule (99.9%), which could be interesting for further chemical processes. Productivity is also high with 123 g L^{-1} starting from 120 g L^{-1} of inulin treated by inulinase. The enzymatic treatment yielded inulin, which was used instead of glucose in MRS medium for fermentation [110, 111].

In another example, lactic acid production from fermented orange peels was evaluated by ion-exchange chromatography. The solid fermentations were in mono or coculture, with *L. casei* 2246, *L. plantarum* 285, and *L. paracasei* 4186. This study showed that fermentation resulted in higher lactic acid production with the monoculture *L. casei* 2246 and the coculture *L. casei* 2246 with *L. plantarum* 285. Glucose can also be converted to lactic acid by symbiotic relationship between different lactic acid bacteria. *L. helveticus* is an AA-producing strain (alanine, serine, aspartate, glutamate, aromatic AA, and histidine), whereas *L. delbrueckii* is a lactic-acid-producing strain but produces little of these AAs necessary for its growth. Thus, this co-fermentation optimized the lactic acid yield [104]. Before industrialization of such a process, scale-up has to be demonstrated and downstream processes (purification) to be implemented and considered. However, another technology could also be used for lactic acid production by microorganisms. Indeed, solid-state fermentation was used with cassava bagasse as substrate and *L. delbruechii* as microorganism [103, 112].

3.4 Other applications of active ingredients produced from fermented plant extracts

Another biotechnology application is the production of proteins, peptides, or AA such as GABA. Indeed, plant by-products are sources of different proteins, which can be hydrolyzed during fermentation by *Lactobacillus* species. These microorganisms, especially *L. plantarum,* have developed a proteolytic system to satisfy their nitrogen requirements. The proteolytic activities and protein hydrolysis patterns are very different from one strain to another. The resulting peptides displayed different biological functions such as angiotensin-converting enzyme inhibition, mineral binding, antidiabetic, satiating, immunomodulating, opioid, AO, or antimicrobial activities [12]. The *L. plantarum* LP-9 strain was used to coproduce GABA and lactic acid from agri-residues such as wheat bran, rice bran, corn bran. The results were compared with the use of cassava (starchy food crop), and the production yields were significant and comparable to this control condition [113]. Co-fermentation of Ginseng root and leaf extract always by *L. plantarum* EJ2014 and *B. subtilis,* also showed GABA production [114]. The fermentation of Kimchi by *L. brevis* BJ20 allows the conversion of glutamic acid into GABA. This process is particularly interesting because GABA has an AO activity demonstrated during the study of DPPH scavenging, superoxide scavenging, and xanthine oxidase inhibition tests [115]. Biotechnology also allows production of cosmetic or pharmaceutical products or surfactants. Biosurfactants production was investigated using *L. paracasei* on enzymatically hydrolyzed vineyard pruning waste. This study presented the complete process for this waste valorization using acid hydrolysis, delignification, and enzymatic hydrolysis steps. Authors have demonstrated the impact of the carbon source extraction process on the biosurfactant composition produced by the strain *L. paracasei* A20 [116].

4. Limitations and future challenges

Faced with environmental and societal problems such as pollution, global warming, and overpopulation, crop yields are increasingly challenging to sustain. Moreover, while demand is increasing in developed countries, poor populations are struggling to feed themselves, and undernutrition is high in these countries. This is why the food and agriculture industry must find solutions to meet the needs of all. Among these, better use of plant resources and better exploitation of their by-products are two solutions of interest. In addition, consumers are looking for more natural and healthy products, and industrials are looking for economically viable bio-based solutions. Fruit and vegetable waste and cereal by-products are likely to be reused because of their quantity and richness in nutrients and bacterial strains suitable for lactic fermentation.

When lactic acid bacteria ferment the nutrients in them, the functional and nutritional properties increase, representing significant opportunities for the agri-food, biotechnology, medical, nutraceutical, and cosmetic industries. As presented in this chapter, the fermentation of plant products by *Lactobacillus* allows the production of numerous bioactive molecules for the development of many applications. Nevertheless, to meet the demand, lactic acid fermentation by *Lactobacillus* requires optimization. First of all, the use of plant by-products requires a crucial design of the fermentation process according to the raw material (solid, liquid, semiliquid fermentations). This design could lead to the development and emergence of new processes that should be able to meet industrial viability, economic returns, and consumer needs. Therefore, much work is still needed on these processes to increase the commercialization of new bio-based products from plant by-products. On the other hand, *Lactobacillus* strains are fastidious bacteria in their nutritional requirements and are not necessarily well adapted metabolically for growth from any substrate, and the use of GMOs is a very limiting criterion for many applications (food, cosmetics, etc.). The growth parameters and enzymatic activities of *Lactobacillus* strains have a major impact on applications, particularly when the fermentation substrate is complex. It is therefore necessary to work on the culture conditions and metabolic adaptation of these strains in order to maximize the enzymatic activities and production rates of the molecules of interest. Therefore, many constraints exist, such as the lack of scientific data and hindsight, the control of culture conditions, and the separation and purification processes to recover bioactive compounds. Further efforts are urgently needed to overcome these problems. Nevertheless, one of the advantages of production with *Lactobacillus* is its ability to produce several types of molecules simultaneously, typically lactic acid and other molecules (derived or transformed from the substrate), which makes the fermentation process industrially interesting. Such multiproduct strategies should be promoted in the near future up to industrial scale.

5. Conclusion

Lactic acid fermentation is an ancestral process performed by numerous bacterial strains. Fermentation conditions, substrates, and potential additives represent challenges and constraints for yield optimization, process stabilization, and standardization. Indeed, lactic fermentation by *Lactobacillus* allows the production of many molecules of interest. When these bacteria ferment plant products, they induce biochemical conversions and the production of phenolic compounds, organic acids, and vitamins through their enzymatic activities. This review highlights the different applications related to the production of these compounds. The latter have bioactivities such as AO, AI, prebiotic, antimicrobial, and many others.

In addition, they are of growing interest to the food industry for their ability to increase nutritional value but also for their use as preservatives and modifiers of organoleptic properties. The different studies reviewed here are looking for alternatives to meet environmental and social consumer demand. In order to reduce production costs and the carbon footprint of the process, genetic engineering and the revalorization of plant by-products appear to be interesting avenues of research to improve the yield of compounds of interest. However, there is still a lack of scientific data on the control of fermentation by *Lactobacillus.* Further studies are needed to identify the biochemical reactions and metabolism of *Lactobacillus* involved in the production of bioactive compounds. In addition, studies are needed to further investigate the mechanisms involved in the bioactivities of interest.

Acknowledgements

The authors would like to thank the financial support for the Charles Viollette Industrial Chair provided by the Lille European Metropolis (MEL) and the Isite of the University of Lille.

Acronyms and abbreviations

AA	Amino acids
ABTS	Acid 2,2'-azino-bis(3-éthylbenzothiazoline-6-sulphonique)
AI	Anti inflammatory
AM	Anti-microbial
AO	Antioxidant
BF	Batch fermentation
Co	Coculture
DNA	Deoxyribonucleic acid
DPPH	2,2-DiPhenyl-1-PicrylHydrazyl
EPS	Exopolysaccharides
FP	Fermentation process
FRAP	Ferric reducing antioxydant power
GABA	Gamma-aminobutyric acid
IL-6, Il-10	Interleukins
IM	Immuno-modulatory
LA	Lactic acid
LF	Liquid fermentation

LPS	Lipopolysaccharide
NO	Nitric oxide
OA	Organic acids
ORAC	Oxygen radical absorbance capacity
PC	Phenolic compounds
PB	Prebiotic effect
ROS	Reactive oxygen species
SBF	Solid batch fermentation
SF	Solid fermentation
SFCA	Short-chain fatty acid
TNF-alpha	Tumor-necrosis factor-alpha

Author details

Morgan Le Rouzic[1,2†], Pauline Bruniaux[1,3†], Cyril Raveschot[3], François Krier[1], Vincent Phalip[1], Rozenn Ravallec[1], Benoit Cudennec[1] and François Coutte[1*]

1 Université de Lille, UMRt BioEcoAgro 1158-INRAE, Institut Charles Viollette, France

2 Eurabiotech, Loos, France

3 VF-Bioscience, Loos, France

*Address all correspondence to: francois.coutte@univ-lille.fr

† Both authors have contributed equally to this work.

References

[1] Lugtenberg BJJ, Malfanova N, Kamilova F, Berg G. Plant growth promotion by Microbes. Molecular Microbial Ecology of the Rhizosphere. 2013;559-573

[2] Hérault B. La demande alimentaire en 2050 : chiffres, incertitudes et marges de manœuvre. Ministère de l'Agriculture, de l'Alimentation, de la Pêche, de la Ruralité et de l'Aménagement du territoire; 2011. Report No.: 27

[3] Department of Economic and Social Affairs, Sustainable Development. THE 17 GOALS | Sustainable Development [Internet]. Organisation des Nations Unies. Available from: https://sdgs.un.org/fr/goals

[4] Septembre-Malaterre A, Remize F, Poucheret P. Fruits and vegetables, as a source of nutritional compounds and phytochemicals: Changes in bioactive compounds during lactic fermentation. Food Research International. 2018;**104**: 86-99

[5] Dogan K, Tornuk F. Improvement of bioavailability of bioactive compounds of medicinal herbs by drying and fermentation with Lactobacillus plantarum. Functional Foods in Health and Disease. 2019;**9**(12):735-748

[6] Ravindran R, Jaiswal AK. Exploitation of food industry waste for high-value products. Trends in Biotechnology. 2016;**34**(1):58-69

[7] Görgüç A, Gençdağ E, Yılmaz FM. Bioactive peptides derived from plant origin by-products: Biological activities and techno-functional utilizations in food developments—A review. Food Research International. 2020;**136**:109504

[8] Bustamante D, Tortajada M, Ramón D, Rojas A. Production of D-lactic acid by the fermentation of orange peel waste hydrolysate by lactic acid bacteria. Fermentation. 2020;**6**(1):1

[9] Wang J, Jiang L, Sun H. Early evidence for beer drinking in a 9000-year-old platform mound in southern China. PLoS One. 2021;**16**(8):e0255833

[10] Capurso L. Thirty Years of Lactobacillus rhamnosus GG: A Review. Journal of Clinical Gastroenterology. 2019;**53**(Supplement 1):S1-S41

[11] Hill D, Sugrue I, Tobin C, Hill C, Stanton C, Ross RP. The lactobacillus casei group: History and health related applications. Frontiers in Microbiology. 2018;**9**:2107

[12] Raveschot C, Cudennec B, Coutte F, Flahaut C, Fremont M, Drider D, et al. Production of bioactive peptides by lactobacillus species: From gene to application. Frontiers in Microbiology. 2018;**9**:2354

[13] Zheng J, Wittouck S, Salvetti E, Franz CMAP, Harris HMB, Mattarelli P, et al. A taxonomic note on the genus Lactobacillus: Description of 23 novel genera, emended description of the genus Lactobacillus Beijerinck 1901, and union of Lactobacillaceae and Leuconostocaceae. International Journal of Systematic and Evolutionary Microbiology. 2020;**70**(4):2782-2858

[14] Muñoz R, de las Rivas B, de Felipe FL, Reverón I, Santamaría L, Esteban-Torres M, et al. Biotransformation of phenolics by lactobacillus plantarum in fermented foods. In Fermented Foods in Health and Disease Prevention. 2017. pp. 63-83

[15] Carpinelli Macedo JV, de Barros Ranke FF, Escaramboni B, Campioni TS, Fernández Núñez EG, de Oliva Neto P. Cost-effective lactic acid production by fermentation of agro-industrial residues. Biocatalysis and Agricultural Biotechnology. 2020;**27**:101706

[16] Hur SJ, Lee SY, Kim Y-C, Choi I, Kim G-B. Effect of fermentation on the antioxidant activity in plant-based foods. Food Chemistry. 2014;**160**:346-356

[17] Ruiz Rodríguez LG, Zamora Gasga VM, Pescuma M, Van Nieuwenhove C, Mozzi F, Sánchez Burgos JA. Fruits and fruit by-products as sources of bioactive compounds. Benefits and trends of lactic acid fermentation in the development of novel fruit-based functional beverages. Food Research International. 2021;**140**:109854

[18] Bintsis T. Department of Agricultural Technology, TEI of West Macedonia, 53100 Florina, Greece. Lactic acid bacteria as starter cultures: An update in their metabolism and genetics. AIMS Microbiology. 2018;**4**(4):665-684

[19] Landete JM, Rodriguez H, De Las RB, Munoz R. High-added-value antioxidants obtained from the degradation of wine phenolics by lactobacillus plantarum. Journal of Food Protection. 2007;**70**(11):2670-2675

[20] Zhou Y, Wang R, Zhang Y, Yang Y, Sun X, Zhang Q, et al. Biotransformation of phenolics and metabolites and the change in antioxidant activity in kiwifruit induced by Lactobacillus plantarum fermentation. Journal of the Science of Food and Agriculture. 2020;**100**(8):3283-3290

[21] Kagkli DM, Corich V, Bovo B, Lante A, Giacomini A. Antiradical and antimicrobial properties of fermented red chicory (Cichorium intybus L.) by-products. Annals of Microbiology. 2016;**66**(4):1377-1386

[22] Leonard W, Zhang P, Ying D, Adhikari B, Fang Z. Fermentation transforms the phenolic profiles and bioactivities of plant-based foods. Biotechnology Advances. 2021;**49**:107763

[23] Feng Y, Zhang M, Mujumdar AS, Gao Z. Turmeric Extract: Potential Use as a Prebiotic and Anti-Inflammatory Compound? Trends in Food Science & Technology. 2017;**65**:40-48

[24] Pianpumepong P, Anal AK, Doungchawee G, Noomhorm A. Study on enhanced absorption of phenolic compounds of Lactobacillus-fermented turmeric (Curcuma longa Linn.) beverages in rats. International Journal of Food Science & Technology. 2012;**47**(11): 2380-2387

[25] Singhal B, Pundhir A, Maurya A. In vitro evaluation of functional attributes of LABs for the development of turmeric based probiotic beverage. American Journal of Research Communication. 2016;**4**:141-173

[26] Wang M, Lei M, Samina N, Chen L, Liu C, Yin T, et al. Impact of Lactobacillus plantarum 423 fermentation on the antioxidant activity and flavor properties of rice bran and wheat bran. Food Chemistry. 2020;**330**:127156

[27] Mukherjee R, Chakraborty R, Dutta A. Comparison of optimization approaches (response surface methodology and artificial neural network-genetic algorithm) for a novel mixed culture approach in soybean meal fermentation. Journal of Food Process Engineering. 2019;**42**(5):e13124

[28] Cheng Y, Wu T, Chu X, Tang S, Cao W, Liang F, et al. Fermented

blueberry pomace with antioxidant properties improves fecal microbiota community structure and short chain fatty acids production in an in vitro mode. LWT. 2020;**125**:109260

[29] Kwaw E, Ma Y, Tchabo W, Apaliya MT, Wu M, Sackey AS, et al. Effect of lactobacillus strains on phenolic profile, color attributes and antioxidant activities of lactic-acid-fermented mulberry juice. Food Chemistry. 2018;**250**:148-154

[30] Lizardo RCM, Cho HD, Won YS, Seo KI. Fermentation with mono- and mixed cultures of Lactobacillus plantarum and L. casei enhances the phytochemical content and biological activities of cherry silverberry (Elaeagnus multiflora Thunb.) fruit. Journal of the Science of Food and Agriculture. 2020;**100**(9):3687-3696

[31] Frediansyah A, Romadhoni F, Nurhayati R, Wibowo AT. Fermentation of jamaican cherries juice using lactobacillus plantarum elevates antioxidant potential and inhibitory activity against type II diabetes-related enzymes. Molecules. 2021;**26**(10):2868

[32] Goto M, Kuda T, Shikano A, Charrouf Z, Yamauchi K, Yokozawa M, et al. Induction of superoxide anion radical-scavenging c. LWT. 2019;**100**:56-61

[33] Jung J, Jang HJ, Eom SJ, Choi NS, Lee N-K, Paik H-D. Fermentation of red ginseng extract by the probiotic Lactobacillus plantarum KCCM 11613P: ginsenoside conversion and antioxidant effects. Journal of Ginseng Research. 2019;**43**(1):20-26

[34] Marazza JA, Nazareno MA, de Giori GS, Garro MS. Enhancement of the antioxidant capacity of soymilk by fermentation with Lactobacillus rhamnosus. Journal of Functional Foods. 2012;**4**(3):594-601

[35] Li Z, Teng J, Lyu Y, Hu X, Zhao Y, Wang M. Enhanced antioxidant activity for apple juice fermented with lactobacillus plantarum ATCC14917. Molecules. 2019;**24**(1):51

[36] Li H, Huang J, Wang Y, Wang X, Ren Y, Yue T, et al. Study on the nutritional characteristics and antioxidant activity of dealcoholized sequentially fermented apple juice with Saccharomyces cerevisiae and Lactobacillus plantarum fermentation. Food Chemistry. 2021;**363**:130351

[37] Gao H, Wen J-J, Hu J-L, Nie Q-X, Chen H-H, Nie S-P, et al. Momordica charantia juice with Lactobacillus plantarum fermentation: Chemical composition, antioxidant properties and aroma profile. Food Bioscience. 2019;**29**:62-72

[38] Wu S-C, Su Y-S, Cheng H-Y. Antioxidant properties of Lactobacillus-fermented and non-fermented Graptopetalum paraguayense E. Walther at different stages of maturity. Food Chemistry. 2011;**129**(3):804-809

[39] Reque PM, Pinilla CMB, Tinello F, Corich V, Lante A, Giacomini A, et al. Biochemical and functional properties of wheat middlings bioprocessed by lactic acid bacteria. Journal of Food Biochemistry. 2020;**44**(7):e13262

[40] Tlais AZA, Da Ros A, Filannino P, Vincentini O, Gobbetti M, Di Cagno R. Biotechnological re-cycling of apple by-products: A reservoir model to produce a dietary supplement fortified with biogenic phenolic compounds. Food Chemistry. 2021;**336**:127616

[41] Jin X, Chen W, Chen H, Chen W, Zhong Q. Combination of Lactobacillus

plantarum and Saccharomyces cerevisiae DV10 as Starter Culture to Produce Mango Slurry: Microbiological. Chemical Parameters and Antioxidant Activity. Molecules. 2019;**24**(23):4349

[42] Mousavi ZE, Mousavi M. The effect of fermentation by Lactobacillus plantarum on the physicochemical and functional properties of liquorice root extract. LWT. 2019;**105**:164-168

[43] Braga ARC, de Mesquita LMS, PLG M, Habu S, de Rosso VV. Lactobacillus fermentation of jussara pulp leads to the enzymatic conversion of anthocyanins increasing antioxidant activity. Journal of Food Composition and Analysis. 2018;**69**:162-170

[44] Vieira ADS, Battistini C, Bedani R, Saad SMI. Acerola by-product may improve the in vitro gastrointestinal resistance of probiotic strains in a plant-based fermented beverage. LWT. 2021;**141**:110858

[45] Thompson HO, Önning G, Holmgren K, Strandler HS, Hultberg M. Fermentation of Cauliflower and White Beans with Lactobacillus plantarum—Impact on Levels of Riboflavin, Folate, Vitamin B12, and Amino Acid Composition. Plant Foods for Human Nutrition. 2020;**75**(2):236-242

[46] Rizzello CG, Nionelli L, Coda R, De Angelis M, Gobbetti M. Effect of sourdough fermentation on stabilisation, and chemical and nutritional characteristics of wheat germ. Food Chemistry. 2010;**119**(3):1079-1089

[47] Nancib A, Nancib N, Meziane-Cherif D, Boubendir A, Fick M, Boudrant J. Joint effect of nitrogen sources and B vitamin supplementation of date juice on lactic acid production by Lactobacillus casei subsp. rhamnosus. Bioresource Technology. 2005;**96**(1):63-67

[48] Chauhan K, Trivedi U, Patel KC. Statistical screening of medium components by Plackett–Burman design for lactic acid production by Lactobacillus sp. KCP01 using date juice. Bioresource Technology. 2007;**98**(1):98-103

[49] Bahry H, Abdalla R, Pons A, Taha S, Vial C. Optimization of lactic acid production using immobilized Lactobacillus Rhamnosus and carob pod waste from the Lebanese food industry. Journal of Biotechnology. 2019;**306**:81-88

[50] Ngouénam JR, Momo Kenfack CH, Foko Kouam EM, Kaktcham PM, Maharjan R, Ngoufack FZ. Lactic acid production ability of Lactobacillus sp. from four tropical fruits using their by-products as carbon source. Heliyon. 2021;**7**(5):e07079

[51] Kim J, Choi K-B, Park JH, Kim KH. Metabolite profile changes and increased antioxidative and antiinflammatory activities of mixed vegetables after fermentation by Lactobacillus plantarum. PLOS ONE. 2019;**14**(5):e0217180

[52] Shahbazi R, Sharifzad F, Bagheri R, Alsadi N, Yasavoli-Sharahi H, Matar C. Anti-inflammatory and immunomodulatory properties of fermented plant foods. Nutrients. 2021;**13**(5):1516

[53] Chon H, Kim G, Kim S. Comparison of aqueous plant extracts before and after fermentation with lactobacillus paracasei LS-2 on cytokine induction and antioxidant activity. Natural Product Communications. 2010;**5**(8):1934578X1000500827

[54] Ali MS, Lee E-B, Lee S-J, Lee S-P, Boby N, Suk K, et al. Aronia melanocarpa extract fermented by lactobacillus plantarum EJ2014 modulates immune response in mice. Antioxidants. 2021;**10**(8):1276

[55] Kim K, Lee G, Thanh HD, Kim J-H, Konkit M, Yoon S, et al. Exopolysaccharide from Lactobacillus plantarum LRCC5310 offers protection against rotavirus-induced diarrhea and regulates inflammatory response. Journal of Dairy Science. 2018;**101**(7):5702-5712

[56] Kim S-B, Kang B-H, Kwon H-S, Kang J-H. Antiinflammatory and antiallergic activity of fermented turmeric by lactobacillus johnsonii IDCC 9203. Microbiology and Biotechnology Letters. 2011;**39**(3):266-273

[57] Ghiamati Yazdi F, Soleimanian-Zad S, van den Worm E, Folkerts G. Turmeric extract: Potential use as a prebiotic and anti-inflammatory compound? Plant Foods Hum Nutr. 2019;**74**(3):293-299

[58] Yong CC, Yoon Y, Yoo HS, Oh S. Effect of lactobacillus fermentation on the anti-inflammatory potential of turmeric. Journal of Microbiology and Biotechnology. 2019;**29**(10):1561-1569

[59] Ayivi RD, Gyawali R, Krastanov A, Aljaloud SO, Worku M, Tahergorabi R, et al. Lactic acid bacteria: Food safety and human health applications. Dairy. 2020;**1**(3):202-232

[60] Cotârleț M, Stănciuc N, Bahrim GE. Yarrowia lipolytica and lactobacillus paracasei solid state fermentation as a valuable biotechnological tool for the Pork Lard and Okara's biotransformation. Microorganisms. 2020;**8**(8):1098

[61] Kothari D, Lee W-D, Jung ES, Niu K-M, Lee CH, Kim S-K. Controlled fermentation using autochthonous lactobacillus plantarum improves antimicrobial potential of chinese chives against poultry pathogens. Antibiotics. 2020;**9**(7):386

[62] Dallagnol AM, Pescuma M, De Valdez GF, Rollán G. Fermentation of quinoa and wheat slurries by Lactobacillus plantarum CRL 778: Proteolytic activity. Applied Microbiology and Biotechnology. 2013;**97**(7):3129-3140

[63] Hashemi SMB, Mousavi Khaneghah A, Barba FJ, Nemati Z, Sohrabi Shokofti S, Alizadeh F. Fermented sweet lemon juice (Citrus limetta) using Lactobacillus plantarum LS5: Chemical composition, antioxidant and antibacterial activities. Journal of Functional Foods. 2017;**38**:409-414

[64] Beck MR, Garrett K, Marshall CJ, Olejar K, Bunt CR, Maxwell TMR, et al. Lactobacillus fermented plant extracts provided to yearling ewes improves their lambs' antioxidant status at weaning. Animal Feed Science and Technology. 2021;**281**:115103

[65] Uerlings J, Schroyen M, Bautil A, Courtin C, Richel A, Sureda EA, et al. In vitro prebiotic potential of agricultural by-products on intestinal fermentation, gut barrier and inflammatory status of piglets. British Journal of Nutrition. 2020;**123**(3):293-307

[66] Zhang Z-P, Ma J, He Y-Y, Lu J, Ren D-F. Antioxidant and hypoglycemic effects of diospyros lotus fruit fermented with Microbacterium flavum and Lactobacillus plantarum. Journal of Bioscience and Bioengineering. 2018;**125**(6):682-687

[67] Park S, Son H-K, Chang H-C, Lee J-J. Effects of cabbage-apple juice fermented by lactobacillus plantarum EM on lipid profile improvement and obesity amelioration in rats. Nutrients. 2020;**12**(4):1135

[68] Ryu J-Y, Kang HR, Cho SK. Changes over the fermentation period in phenolic compounds and antioxidant and anticancer activities of blueberries fermented by lactobacillus

plantarum. Journal of Food Science. 2019;**84**(8):2347-2356

[69] Pontonio E, Dingeo C, Gobbetti M, Rizzello CG. Maize milling by-products: From food wastes to functional ingredients through lactic acid bacteria fermentation. Frontiers in Microbiology. 2019;**10**:561

[70] Pontonio E, Dingeo C, Di Cagno R, Blandino M, Gobbetti M, Rizzello CG. Brans from hull-less barley, emmer and pigmented wheat varieties: From by-products to bread nutritional improvers using selected lactic acid bacteria and xylanase. International Journal of Food Microbiology. 2020;**313**:108384

[71] Chen H-Y, Hsieh C-W, Chen P-C, Lin S-P, Lin Y-F, Cheng K-C. Development and optimization of djulis sourdough bread fermented by lactic acid bacteria for antioxidant capacity. Molecules. 2021;**26**(18):5658

[72] Schettino R, Pontonio E, Rizzello CG. Use of fermented hemp, chickpea and milling by-products to improve the nutritional value of semolina pasta. Foods. 2019;**8**(12):604

[73] Brückner-Gühmann M, Banovic M, Drusch S. Towards an increased plant protein intake: Rheological properties, sensory perception and consumer acceptability of lactic acid fermented, oat-based gels. Food Hydrocolloids. 2019;**96**:201-208

[74] Kachouri F, Hamdi M. Use lactobacillus plantarum in olive oil process and improvement of phenolic compounds content. Journal of Food Engineering. 2006;**77**(3):746-752

[75] Ruiz-Barba J, Brenes-Balbuena M, Jiménez-Díaz R, García-García P, Garrido-Fernández A. Inhibition of Lactobacillus plantarum by polyphenols extracted from two different kinds of olive brine. Journal of Applied Bacteriology. 1993;**74**(1):15-19

[76] Leal-Sánchez MV, Ruiz-Barba JL, Sánchez AH, Rejano L, Jiménez-Díaz R, Garrido A. Fermentation profile and optimization of green olive fermentation using lactobacillus plantarum LPCO10 as a starter culture. Food Microbiology. 2003;**20**(4):421-430

[77] Landete JM, Curiel JA, Rodríguez H, de las Rivas B, Muñoz R. Study of the inhibitory activity of phenolic compounds found in olive products and their degradation by Lactobacillus plantarum strains. Food Chemistry. 2008;**107**(1):320-326

[78] Ciafardini G, Marsilio V, Lanza B, Pozzi N. Hydrolysis of Oleuropein by *Lactobacillus plantarum* Strains Associated with Olive Fermentation. Applied and Environmental Microbiology. 1994;**60**(11):4142-4147

[79] Ruiz-Barba JL, Cathcart DP, Warner PJ, Jiménez-Díaz R. Use of Lactobacillus plantaru LPCO10, a Bacteriocin Producer, as a Starter Culture in Spanish-Style Green Olive Fermentations. Applied and Environmental Microbiology. 1994;**60**(6):2059-2064

[80] Kachouri F, Setti K, Ksontini H, Mechmeche M, Hamdi M. Improvement of antioxidant activity of olive mill wastewater phenolic compounds by Lactobacillus plantarum fermentation. Desalination and Water Treatment. 2016;**57**(56):27125-27137

[81] Tangyu M, Muller J, Bolten CJ, Wittmann C. Fermentation of plant-based milk alternatives for improved flavour and nutritional value. Applied and Microbiology Biotechnology. 2019;**103**(23-24):9263-9275

[82] Do TVT, Fan L. Probiotic viability, qualitative characteristics, and

sensory acceptability of vegetable juice mixture fermented with lactobacillus strains. Food and Nutrition Sciences. 2019;**10**(04):412

[83] Cui S, Zhao N, Lu W, Zhao F, Zheng S, Wang W, et al. Effect of different Lactobacillus species on volatile and nonvolatile flavor compounds in juices fermentation. Food Science & Nutrition. 2019;**7**(7):2214-2223

[84] Bianchi F, Rossi EA, Gomes RG, Sivieri K. Potentially synbiotic fermented beverage with aqueous extracts of quinoa (Chenopodium quinoa Willd) and soy. Food Science and Technology International. 2015;**21**(6):403-415

[85] Yoon KY, Woodams EE, Hang YD. Production of probiotic cabbage juice by lactic acid bacteria. Bioresource Technology. 2006;**97**(12):1427-1430

[86] Lavinia BC, Manea I, Bratu MG, Avram D, Nicolescu CL. Evaluation of the cabbage and cucumber juices as substrate for Lactobacillus acidophilus LA-5. Romanian Biotechnological Letters. 2012;**17**(4):12

[87] Karovičová J, Drdák M, Greif G, Hybenová E. The choice of strains of Lactobacillus species for the lactic acid fermentation of vegetable juices. European Food Research and Technology. 1999;**210**(1):53-56

[88] Jang H, Kim M. Characteristics of vegetable juice fermented with lactobacillus plantarum MKHA15 and leuconostoc mesenteroids MKSR. Journal of the Korean Dietetic Association. 2019;**25**(4):281-294

[89] Dunkley KE, Hekmat S. Development of probiotic vegetable juice using Lactobacillus Rhamnosus GR-1. Nutrition & Food Science. 2020;**50**(5):955-968

[90] Hashemi SMB, Jafarpour D. Fermentation of bergamot juice with Lactobacillus plantarum strains in pure and mixed fermentations: Chemical composition, antioxidant activity and sensorial properties. LWT. 2020;**131**: 109803

[91] Gupta S, Jaiswal AK, Abu-Ghannam N. Optimization of fermentation conditions for the utilization of brewing waste to develop a nutraceutical rich liquid product. Industrial Crops and Products. 2013;**44**:272-282

[92] Oktaviani L, Astuti DI, Rosmiati M, Abduh MY. Fermentation of coffee pulp using indigenous lactic acid bacteria with simultaneous aeration to produce cascara with a high antioxidant activity. Heliyon. 2020;**6**(7):e04462

[93] Nguyen NK, Dong NTN, Nguyen HT, Le PH. Lactic acid bacteria: promising supplements for enhancing the biological activities of kombucha. SpringerPlus. 2015;**4**(1):91

[94] Nguyen NK, Dong NTN, Le PH, Nguyen HT. Evaluation of the glucuronic acid production and other biological activities of fermented sweeten-black tea by kombucha layer and the co-culture with different lactobacillus sp. strains. International Journal Of Modern Engineering Research. 2014;**4**(5):12-17

[95] Hou J, Luo R, Ni H, Li K, Mgomi FC, Fan L, et al. Antimicrobial potential of kombucha against foodborne pathogens: A review. qas. 2021;**13**(3):53-61

[96] Park JH, Kim Y, Kim SH. Green tea extract (camellia sinensis) fermented by lactobacillus fermentum attenuates alcohol-induced liver damage. Bioscience, Biotechnology, and Biochemistry. 2012;**76**(12):2294-2300

[97] Cherdthong A, Suntara C, Khota W. Lactobacillus casei TH14 and additives

could modulate the quality, gas kinetics and the in vitro digestibility of ensilaged rice straw. Journal of Animal Physiology and Animal Nutrition. 2020;**104**(6):1690-1703

[98] Xu Z, He H, Zhang S, Kong J. Effects of inoculants Lactobacillus brevis and Lactobacillus parafarraginis on the fermentation characteristics and microbial communities of corn stover silage. Scientific Reports. 2017;**7**(1):13614

[99] Irsyammawati A, Ndaru PH. The effect of lactobacillus plantarum addition and fermentation periods on nutritive value dwarf elephant grass (pennisetum purpureum cv mott) silage. IOP Conference Series: Earth Environmental Science. 2020;**478**:012049

[100] Zhao J, Dong Z, Li J, Chen L, Bai Y, Jia Y, et al. Evaluation of Lactobacillus plantarum MTD1 and waste molasses as fermentation modifier to increase silage quality and reduce ruminal greenhouse gas emissions of rice straw. Science of The Total Environment. 2019;**688**:143-152

[101] Mu L, Xie Z, Hu L, Chen G, Zhang Z. Lactobacillus plantarum and molasses alter dynamic chemical composition, microbial community, and aerobic stability of mixed (amaranth and rice straw) silage. Journal of the Science of Food and Agriculture. 2021;**101**(12):5225-5235

[102] Risdianto D, Suthama N, Suprijatna E, Sunarso S. Inclusion effect of ginger and turmeric mixture combined with Lactobacillus spp. isolated from rumen fluid of cattle on health status and growth of broiler. Journal of the Indonesian Tropical Animal Agriculture. 2019;**44**(4):423

[103] Dedenaro G, Costa S, Rugiero I, Pedrini P, Tamburini E. Valorization of agri-food waste via fermentation: Production of l-lactic acid as a building block for the synthesis of biopolymers. Applied Sciences. 2016;**6**(12):379

[104] Ricci A, Diaz AB, Caro I, Bernini V, Galaverna G, Lazzi C, et al. Orange peels: From by-product to resource through lactic acid fermentation. Journal of the Science of Food and Agriculture. 2019;**99**(15):6761-6767

[105] Hwang HJ, Lee SY, Kim SM, Lee SB. Fermentation of seaweed sugars by Lactobacillus species and the potential of seaweed as a biomass feedstock. Biotechnol Bioproc E. 2011;**16**(6):1231-1239

[106] Zhang Y, Vadlani PV, Kumar A, Hardwidge PR, Govind R, Tanaka T, et al. Enhanced D-lactic acid production from renewable resources using engineered Lactobacillus plantarum. Applied and Microbiology Biotechnology. 2016;**100**(1):279-288

[107] Zhang Y, Vadlani PV. d-Lactic acid biosynthesis from biomass-derived sugars via Lactobacillus delbrueckii fermentation. Bioprocess and Biosystem Engineering. 2013;**36**(12):1897-1904

[108] Göksungur Y, Güvenç U. Batch and continuous production of lactic acid from beet molasses by lactobacillus delbrueckii IFO 3202. Journal of Chemical Technology & Biotechnology. 1997;**69**(4):399-404

[109] Li Z, Ding S, Li Z, Tan T. L-lactic acid production by Lactobacillus casei fermentation with corn steep liquor-supplemented acid-hydrolysate of soybean meal. Biotechnology Journal. 2006;**1**(12):1453-1458

[110] Tu W-L, Hsu T-C, Wang C-A, Guo G-L, Chao Y. Using novel lactobacillus plantarum to produce lactic acid from lignocellulosic biomass in an integrated

simultaneous saccharification and fermentation process. BioResources. 2019;**14**(2):3873-3885

[111] Xu Q, Zang Y, Zhou J, Liu P, Li X, Yong Q, et al. Highly efficient production of d-lactic acid from chicory-derived inulin by Lactobacillus bulgaricus. Bioprocess and Biosystem Engineering. 2016;**39**(11):1749-1757

[112] John RP, Nampoothiri KM, Pandey A. Solid-state fermentation for l-lactic acid product. Process Biochemistry. 2006;**41**(4):759-763

[113] Sharma P, Sharma A, Singh J, Singh N, Singh S, Tomar GS, et al. Co-production of gamma amino butyric acid (GABA) and lactic acid using Lactobacillus plantarum LP-9 from agro-residues. Environmental Technology & Innovation. 2021;**23**:101650

[114] Park J-H, Garcia CV, Lee S-P. Fortification of poly-γ-glutamic acid and γ-aminobutyric acid in homogenized hydroponic ginseng co-fermented by bacillus subtilis ha and lactobacillus plantarum EJ2014. Preventive Nutrition and Food Science. 2019;**24**(4):485-491

[115] Lee B-J, Kim J-S, Kang YM, Lim J-H, Kim Y-M, Lee M-S, et al. Antioxidant activity and γ-aminobutyric acid (GABA) content in sea tangle fermented by Lactobacillus brevis BJ20 isolated from traditional fermented foods. Food Chemistry. 2010;**122**(1):271-276

[116] Vecino X, Rodríguez-López L, Gudiña EJ, Cruz JM, Moldes AB, Rodrigues LR. Vineyard pruning waste as an alternative carbon source to produce novel biosurfactants by Lactobacillus paracasei. Journal of Industrial and Engineering Chemistry. 2017;**55**:40-49

Chapter 3

Diverse Bioactive Molecules from the Genus *Lactobacillus*

Rodney H. Perez and Amily E. Ancuelo

Abstract

Lactobacilli are widespread microorganisms and are broadly employed in a variety of applications. It is one of the LAB genera that has been designated as Generally Regarded as Safe (GRAS) and many of its member species are included in the Qualified Presumption of Safety (QPS) list. *Lactobacillus* is commonly utilized as a starter culture in many fermented food products, probiotics, and has long been used as natural bio-preservatives to increase shelf life and improve food quality and safety. Aside from the many benefits, it delivers in the food sector, the use of *lactobacillus* strains in the clinical setting as a prophylactic and/or treatment for a variety of diseases has gained increasing attention. These uses of *lactobacillus* are all made possible through the diverse bioactive molecules it generates. *Lactobacillus* exerts its positive health and nutritional effects through a variety of mechanisms, including inhibition of pathogen adhesion or colonization, metabolic activity through the synthesis of metabolites and enzymes, and immune system modulation among others. The ability of many *lactobacillus* strains to mediate the bio-conversion of certain metabolites has also been shown in numerous studies. This chapter describes the recent findings on the impact of the diverse bioactive molecules produced by different *lactobacillus* strains, their mode of action, and their application in different industries.

Keywords: lactic acid bacteria, GRAS, *lactobacillus*, bioactive compounds, probiotics

1. Introduction

One of the most significant, and extensively used lactic acid bacteria (LAB) is *lactobacillus*. This genus comprises a large number of species that can be found in diverse environments, such as in plants, food products, and mucosal surfaces of the human body. Lactobacilli are characterized by the formation of lactic acid as the primary end product of carbohydrate metabolism. Species of *lactobacillus* are Gram-positive, homofermentative, thermophilic, and non-spore-forming rods. *Lactobacillus* species ferment a broad array of carbohydrates and can ferment extracellular fructans, starch, or glycogen depending on the strain. All organisms formerly assigned to the *Lactobacillus delbrueckii* group are now included in the emended description of the genus [1].

Among LAB, *lactobacillus* is one of the genera considered as Generally Regarded as Safe (GRAS) by the U.S. Food and Drug Administration (FDA) as they are a safe means to generate products for a variety of industries. The European Food Safety

Authority (EFSA) also grants Qualified Presumption of Safety (QPS) status to many *lactobacillus* species. The EFSA notes that some LAB strains are susceptible to acquiring virulence and antibiotic resistance genes and have opportunistic properties and hence excluded in the QPS list [2]. Lactobacilli are broadly applied in the food industry as both technical starters in fermented goods and probiotics due to their unique health and nutritional benefits [3, 4]. These microorganisms play an indispensable role in many foods fermentation. For instance, many lactobacilli species have been identified as primary microorganisms in sauerkraut fermentation [5]. Lactobacilli also play a significant part in malolactic fermentation, critical in winemaking [6]. In Chinese Maotai-flavored liquor production, *lactobacillus* accelerates flavor component conversion from alcohol (ethanol) to acid (lactic acid and acetic acid) [7]. Lactic acid fermentation using *L. plantarum* was found optimal in improving pea protein isolates aroma and taste [8]. Fermentation with *lactobacillus* also improved the foaming capabilities of egg white [9]. Many *lactobacillus* strains are also important in the manufacture of many fermented meat and meat products. Moreover, probiotic *lactobacillus* was given special attention as well for its ability to stimulate or modulate the immune system [10]. It also has possible applications in health-related areas such as intestinal inflammation [11], prevention of urinary tract infection [12], and treatment against cancer cells [13].

Indigenous LAB is constantly exposed to extreme conditions such as varying temperature, pH, and nutrient levels [14]. As a result, native LAB has been linked to higher competitive metabolic capacities, which encourage their growth as a competitive microbiota for other microorganisms in their natural habitat. The synthesis of a large number of bioactive metabolites is one of these capacities. Among LAB genera, lactobacilli are known producers of diverse bioactive molecules that offer a wide range of benefits to food, agricultural, industrial and clinical fields. They have long been exploited in food and animal feed as natural preservatives. Their antimicrobial action is mostly due to the production of organic acids, hydrogen peroxide, inhibitory compounds, as well as competition for nutrients and the development of antimicrobial compounds like bacteriocin [15]. Several studies have shown that the organic acids produced by *lactobacillus* like lactic acid kill pathogens at sufficient concentrations, such as *Campylobacter jejuni* by disrupting its membrane [16]. It also secretes ethanol and fatty acid as antimicrobial molecules. It can produce acetic acid, formic acid, and other acids [17].

Lactobacilli exhibit their beneficial properties through a wide range of processes that include a large spectrum of bioactive compounds. In this chapter, the utility of these microorganisms and their bioactive compound by-products for the promotion of better health and nutrition are summed up. Lactobacilli and their by-products can be utilized in technology and product development geared towards sustainable approaches for the improvement of human conditions targeted by the United Nations 2030 Agenda and its Sustainable Development Goals. The well-established functions of lactobacilli and their bioactive molecules in food fermentation could play a key role in ensuring that people around the world have access to safe and nutritious food by improving the current food production, safety, and preservation. Its application in the clinical setting also has the potential to address major health concerns, including sexual, reproductive, newborn, and environmental diseases which will be discussed in the following sections. Thus, this chapter will center the attention on the different bioactive molecules from the genus *lactobacillus*, such as bacteriocins, bioactive peptides, SCFAs, vitamins, enzymes, EPs, immune-modulating compounds, bioconverted molecules, and its probiotic properties.

2. Bioactive compounds produced by *lactobacillus*

2.1 Bacteriocins

Bacteriocins are multifunctional, ribosomal-synthesized antimicrobial peptides. The bactericidal activity of bacteriocins is demonstrated against species that are closely related to the producer strain [18]. The bactericidal or bacteriostatic actions of bacteriocins produced by Gram-positive bacteria, including LAB, are mostly against Gram-positive bacteria including food-borne pathogens [19]. Bacteriocins inhibit their target cells by destabilizing the bacterial cell membrane and/or creating pores resulting in the death of the target cells through a fast-acting mode of action that is active even at very low concentrations [18]. Bacteriocins from Gram-positive bacteria are divided into three classes based on their structural and physicochemical properties: class I (lantibiotics), which are lanthionine-containing peptides; class II, comprise the non-lanthionine-containing bacteriocins [20].

The promise of bacteriocins, particularly from lactic acid bacteria, for various applications has instigated a great deal of interest in bacteriocin research. LAB bacteriocins are recognized for their activity over a wide pH range and are inherently tolerant to extreme thermal stress. The fact that these antimicrobial peptides are colorless, odorless, and tasteless, adds to their potential uses [18]. Bacteriocins also offer a number of advantages over traditional antibiotics. The most notable of which is that they are primary metabolites with relatively straightforward biosynthetic processes compared to conventional antibiotics, which are secondary metabolites. Thus, bioengineering may readily improve their activity or specificity towards their target bacteria [18].

Lactobacillus, is a LAB genus that has been shown to produce diverse bacteriocins (**Table 1**). A *lactobacillus* strain isolated from traditional Egyptian dairy products showed antimicrobial actions against different Gram-positive and Gram-negative bacteria [35]. The highest inhibitory activity of *L. brevis* (B23) was exhibited against *Escherichia coli, Staphylococcus and Bacillus*.

A *lactobacillus* strain has also been described to produce multiple bacteriocins. *Lactobacillus sakei* 5 from malted barley was found to produce three bacteriocins [36]. Genetic and functional analysis revealed that this strain generates a plasmid-encoded bacteriocin sakacin P, as well as two novels, chromosomally encoded bacteriocins, sakacin T and sakacin X. This strain may be a viable candidate for usage in the brewing sector since it inhibits bacterial strains known to cause severe spoiling problems in this industry.

A strain isolated from human breast milk, *Lactobacillus gasseri* LM19, was also found to produce several bacteriocins, including a novel bacteriocin, gassericin M [28]. In a complex environment that mimicked human colon circumstances, *L. gasseri* LM19 not only survived but also expressed seven bacteriocin genes and generated short-chain fatty acids. The gut origin of *L. gasseri* LM19 enabled it to thrive in GI tract conditions and display antagonistic properties against other gut bacteria, such as enteropathogens [28]. Different bacteriocin-producing *L. plantarum* strains have also been found in a variety of foods, including meat [37], fermented milk [38], cheese [39], and sourdough [40].

Aside from its usefulness in the food industry, bacteriocin-producing *lactobacillus* was also investigated for its use in the clinical setting, particularly in preventing and treating vaginal disorders. *Lactobacillus* is bacteria naturally found in the healthy human vagina [41] and urethra [42]. A low count of *lactobacillus* is

Class	Bacteriocin	*Lactobacillus* strain	Reference
I	Paraplantaricin TC318	*L. paraplantarum* OSY-TC318	[21]
	Plantaricin C	*L. plantarum* LL441	[22]
	Lactocin S	*L. sakei* L45	[23]
IIa	Plantaricin 423	*L. plantarum* 423	[24]
	Plantaricin LPL-1	*L. plantarum* LPL-1	[25]
	Rhamnocin 519	*L. rhamnosus* CJNU 0519	[26]
IIb	Gassericin S	*L. gasseri* LA327	[27]
	Gassericin T		
	Gassericin M	*L. gasseri* LM19	[28]
IIc	Acidocin B	*L. acidophilus* M46	[29]
	Plantaricyclin A	*L. plantarum* NI326	[30]
	Plantacyclin B21AG	*L. plantarum* WCFS1	[31]
IId	Sakacin D98a	*L. sakei* D98	[32]
	Sakacin D98c		
	Bactofencin A	*L. salivarus* DPC6502	[33]
III	Helveticin-M	*L. crispatus*	[34]

Table 1.
Diverse bacteriocins produced by various strains of Lactobacillus.

inversely related to high numbers of *E. coli* in the vagina and a history of recurrent urinary tract infection [43]. A new bacteriocin generated by *L. acidophilus* KS400 was identified and characterized, as well as its antimicrobial properties against urogenital pathogens [44]. These species have been shown to colonize the epithelial surface and release antimicrobial compounds that regulate the vaginal microflora.

2.2 Bioactive peptides

Digestive proteases and peptidases from human's release food-encrypted bioactive peptides that can be absorbed by the gut and then reach peripheral organs. However, the enzymatic activity of LAB largely contributes to their release, either into the food matrix or in the gut. Due to the limited length of the overall genome, the biosynthetic abilities of LAB are very limited especially in amino acid synthesis [45]. Therefore, LAB evolved a complex and sophisticated proteolytic system allowing them to get amino acids from the proteins present in the external environment [46]. The proteolytic system of LAB converts protein substrates into free amino acids and small peptides, which enables them to carry out their intrinsic physiological mechanisms such as regulation of intracellular pH, production of metabolic energy, stress tolerance, and biosynthesis of proteins [47].

Numerous bioactive peptides lack activity when protein is encrypted, but display their interesting biological functions when released proteolytically. They have been shown to hold health-promoting qualities as antimicrobials, hypocholesterolemic, opioid antagonists, angiotensin-converting enzyme inhibitors, anti-thrombotic, immuno-modulators, cytomodulators, and antioxidants [48]. The utilization of LAB such as *lactobacillus* in the synthesis and valorization of new bioactive peptides is a useful method. The proteolytic activity of *lactobacillus* is strain- and species-dependent: each species has a distinct proteinase composition, encompassing a wide range of proteolytic activities [49].

Over the past years, *lactobacillus* species have presented great potential as producers of bioactive peptides through fermentation using different protein matrices (**Table 2**). LAB proteolytic system is capable of producing bioactive peptides from a variety of food proteins, particularly casein, which is the major nitrogen source in their environment. [59]. *L. helveticus* CICC6024 was employed to effectively ferment milk-casein under fixed fermentation processes to facilitate an efficient bioactive peptide synthesis [60].

Other *lactobacillus* strains have also been found to be capable of releasing bioactive peptides from food proteins. Milk inoculated with *L. helveticus* and casein hydrolysates generated by *L. helveticus* CP790 extracellular proteinase both contain antihypertensive peptides [61]. Antihypertensive substances were also recovered from an *L. casei* cell extract used in fermented milk production [50]. Two fermented kinds of milk containing ACE-inhibitory peptides were generated using *L. delbrueckii* subsp. *bulgaricus* and *Lactococcus lactis* subsp. *cremoris* strains [62]. Bioactive peptides have been discovered in UHT milk fermented by the probiotic *Lactobacillus* GG strain and digested by pepsin and trypsin enzymes. These bioactive peptides exhibit varying degrees of immunostimulatory, opioid, and ACE-inhibitory properties [55].

***Lactobacillus* strain**	**Bioactive peptides**	**Protein Source**	**Reference**
L. casei	Antihypertensive peptides	Milk	[50]
L. casei *L. plantarum* *L. rhamnosus*	Antimicrobial, antioxidant and ACE-inhibitory peptides	Whey and skim milk	[51]
L. helveticus	Antioxidative, opioid, stimulating, hypotensive, immunomodulating, antibacterial, and antithrombotic peptides	κ-casein	[52]
L. helveticus	ACE-inhibitory peptides	Milk	[53]
L. helveticus R0389 *L. rhamnosus* R0011	ACE-inhibitory and immunomodulating peptides	Casein	[54]
Lactobacillus GG	Immunostimulatory, opioid, and ACE-inhibitory peptides	UHT Milk	[55]
L. plantarum 55	Anti-inflammatory, antihemolytic, antioxidant, antimutagenic, and antimicrobial peptides	Milk	[56]
L. plantarum C2	Anti-oxidant and ACE-inhibitory peptides	Soy milk	[57]
L. sanfranciscensis I4 *L. farciminis* A19 *L. rossiae* A20	Anti-oxidant and anti-inflammatory peptides	Italian sourdough	[58]

Table 2.
Bioactive peptides generated by Lactobacillus *using different protein matrices.*

Cultures of *L. plantarum*, *L. casei*, and *L. rhamnosus* from a fecal sample of a human infant were employed as a proteolytic starter culture for the fermentation of skim milk and whey to release small peptides that have antimicrobial, antioxidant, and ACE inhibitory activities. These encrypted bioactive peptides can be utilized as a functional food and/or, dietary supplement, to provide particular health advantages.

Single-activity milk-derived bioactive peptides have been widely reported. The anti-inflammatory, antihemolytic, antioxidant, antimutagenic, and antimicrobial activities of crude extracts and peptide fractions obtained from fermented milk with specific *L. plantarum* strains were assessed [56]. *L. plantarum* 55 was found to generate encrypted peptides with extensive capabilities as dietary bioactive components for the development of nutraceutical biotechnological products.

2.3 Short-chain fatty acids (SCFAs)

The small and large intestines of humans lack several carbohydrate-digesting enzymes that can be produced by probiotic bacteria. However, the probiotic bacteria ferment these undigested carbohydrates and produce energy that is utilized by the host to carry out various functions. The undigested sugars are converted into short-chain fatty acids (SCFAs) such as butyrate, acetate, and propionate. The typical reaction of SCFAs production and overall stoichiometry has been summarized and is shown as follows [63]:

$$\begin{aligned} 59\,C_6H_{12}O_6 + 38\,H_2O \rightarrow 60\,CH_3COOH + 18\,CH_3CH_2CH_2COOH \\ + 22\,CH_3CH_2COOH + 96\,CO_2 + 134\,H_2 + \text{Heat} \end{aligned} \tag{1}$$

SCFAs from *lactobacillus* have been proven to have therapeutic effects against several diseases through their antimicrobial potential (**Table 3**). For example, *L. reuteri* produces SCFAs to inhibit colon cancer cell proliferation [69]. Several *L. reuteri* strains were shown to synthesize SCFAs and demonstrated growth inhibitory activity against colorectal cancer cells. Thus, the anti-cancer action and the ability to generate anti-carcinogenic active substances of *L. reuteri* indicate that it may be used as a bio-therapeutic. Moreover, the impact of *L. paracasei* CNCM I-1572 on clinical and gut microbiota-related parameters in irritable bowel syndrome (IBS) was also investigated [65]. *L. paracasei* CNCM I-1572 was shown to regulate the structure and function of gut microbiota and decrease immunological activation in IBS by substantially increasing the SCFAs acetate and butyrate and a corresponding decrease in the pro-inflammatory cytokine interleukin-15. Several *lactobacillus* strains were also investigated for their application to treat bacterial vaginosis [64]. The strain *L. plantarum* ZX27 was found to produce more short-chain fatty acids and lactic acid and inhibited *Gardnerella vaginalis* growth and adherence.

SCFAs are generated by bacteria in the gastrointestinal system, which relies on non-digestible carbohydrates for energy. SCFA production is necessary to increase the acidity of the gut environment, which inhibits many harmful microorganisms. Production of SCFAs has been shown as one mechanism of *lactobacillus* strains to inhibit the development of metabolic syndrome by its influence on microbiota modulation [71]. The production of SCFAs and antimicrobial activity of *L. plantarum* G72 for its potential application in improving the diet of pregnant women.

***Lactobacillus* strain**	**SCFA**	**Spectrum**	**Reference**
L. delbrueckii DM8909	Acetatic acid, propionic acid, butyric acid, isobutyric acid, valeric acid, and isovaleric acid	*Gardnerella vaginalis*	[64]
L. plantarum ATCC14917			
L. plantarum ZX27			
L. paracasei CNCM I-1572	Acetatic acid and butyric acid	*Ruminococcus bromii* and *Ruminococcus callidus*	[65]
L. paracasei subsp. *paracasei* NTU 101	Acetic acid, propionic acid, and butyric acid	*Clostridium perfringens* and *Enterobacteriaceae*	[66]
L. plantarum G72	Formic acid, acetic acid, propionic acid, and butyric acid	*Listeria monocytogenes, Salmonella typhimurium, E. coli*, and *S. aureus*	[67]
L. rhamnosus	Acetic acid, propionic acid	*E. coli*	[68]
L. reuteri NCIMB -11,951, −701,359, −701,089, −702,655, and − 702,656	Acetic acid, propionic acid, and butyric acid	Caco-2 colon cancer cells	[69]
L. salivarius ssp. salicinius JCM 1230	Propionic acid, and butyric acid	*Salmonella*	[70]
L. agilis JCM 1048			

Table 3.
Antimicrobial spectrum of the SCFAs produced by Lactobacillus *strain.*

2.4 Vitamins

Vitamins are essential micronutrients that are required for the metabolism of every organism. Humans are incapable of producing vitamins, resulting in vitamin deficiencies, malnutrition, and stunted growth from infants to the elderly. Thus, they must be acquired exogenously (i.e., in the form of diet). All vitamins can be classified into two groups: water-soluble vitamins and fat-soluble vitamins. Water-soluble B-group vitamins are generated by several bacteria and are consumed in the gut. Fat-soluble vitamins, on the other hand, are taken in the digestive tract using lipids as micelles. Plants and animals are natural providers of vitamins, although certain vitamins are chemically produced.

Lactic acid bacteria, especially *lactobacillus*, are known to be good producers of vitamins. *Lactobacillus* decreases the overall growth of bacteria-caused diseases by generating these nutritional components (**Table 4**). *Lactobacillus* strains from traditional yogurt were able to produce B-group vitamins [73]. *L. paracasei* subsp. *tolerance* JCM 1171 (T), *L. acidophilus* KU, and *L. fermentum* showed the highest amount of Vitamin B_6 and B_9, B_3, and B_2, respectively. *L. plantarum* LZ95 originally from infant feces and CY2 from fresh milk were identified to be capable of producing a high level of extracellular vitamin B_{12} as well [75]. Moreover, co-fermentation of glycerol and fructose in soy-yogurt by *L. reuteri* has been demonstrated to enhance vitamin B_{12} synthesis [77].

Some vitamins, particularly riboflavin and folate derivatives, have been shown to help combat certain diseases. Vitamin-producing lactic acid bacteria, particularly strains that produce folate and riboflavin in combination with immune-stimulating

Lactobacillus strain	Vitamins	Reference
L. fermentum KGL2	Vitamin B_2, B_9, and B_{12}	[72]
L. plantarum KGL3A		
L. fermentum KGL4		
L. rhamnosus RNS4		
L. fermentum WTS4		
L. paracasei subsp. *tolerance* JCM 1171^T	Vitamin B_2, B_3, B_6, and B_9	[73]
L. acidophilus KU		
L. fermentum		
L. plantarum CRL2130	Vitamin B_2	[74]
L. plantarum LZ227	Vitamin B_2 and B_9	
L. plantarum LZ95	Vitamin B_{12} (adenosylcobalamin and methylcobalamin)	[75]
L. plantarum CY2		
L. plantarum BHM10	Vitamin B_{12}	[76]
L. plantarum BCF20		
L. reuteri	Vitamin B_{12}	[77]
L. rossiae DSM 15814^T	Vitamin B_{12}	[78]

Table 4.
Example of Lactobacillus *strains known to produce vitamins.*

strains, could be used as effective alternative types of treatment in patients suffering from a variety of inflammatory diseases [79]. Riboflavin-producing *L. plantarum* CRL2130, through oral administration, exhibited its ability to prevent trinitrobenzene sulfonic acid-induced colitis in mice, reducing pro-inflammatory cytokines [74].

2.5 Enzymes

Lactic acid bacteria perform metabolic processes due to the synthesis of enzymes. Enzymes play a critical role in biological reactions by acting as biocatalysts, mediating all anabolic and catabolic pathways, and lowering the activation energy of biochemical reactions. The digestive enzymes in the lysosomes, for example, enhance the digestion of a wide range of substances absorbed from outside the cell in the gastrointestinal tract (GIT). These enzymes work together to convert carbohydrates, proteins, and lipids into monomers that can be absorbed by human cells. Examples of digestive enzymes include amylase, lactase, pepsin, trypsin, pancreatic amylase, lipase, nuclease, maltase, and lactase [80].

Lactobacillus strains are well-known enzyme producers (**Table 5**). Amylases are one of the most often utilized enzymes in industry. These enzymes hydrolyze starch molecules into polymers made up of glucose units [90]. *Lactobacillus* amylases are considered safe since they are non-pathogenic and the end product of their fermentation is lactate, a commonly utilized flavoring ingredient in the food industry [91]. Several *lactobacillus* strains such as *L. brevis, L. casei*, and *L. fermentum*, were shown to produce a significant quantity of amylase [81]. The amylolytic potential of

***Lactobacillus* strain**	**Enzymes**	**Reference**
L. brevis	Amylase	[81]
L. fermentum		
L. brevis C10CpSA3b6	β-glucanase	[82]
L. crispatus I12pSA3b6		
L. brevis I23pSA3b6		
L. fermentum I25pSA3b6		
L. brevis I211pSA3b6		
L. brevis I218pSA3b6		
L. bulgaricus	β-galactosidase enzyme	[83]
L. casei	Amylase and invertase	[81]
L. casei LFTI® L26	ACE-inhibitory enzyme	[84]
Lactobacillus delbrueckii QS306	ACE-inhibitory enzyme	[85]
L. helveticus IMAU80872, IMAU80852, and IMAU80851	ACE-inhibitory enzyme	[86]
L. plantarum	α-galactosidase enzyme	[87]
Lactobacillus rhamnosus	β-galactosidase enzyme	[88]
Lactobacillus sp. G3_4_1TO2	Amylase	[89]

Table 5.
Example of enzyme-producing Lactobacillus *strains.*

lactobacillus strains from wet-milled cereals, cassava flour, and fruits has been studied. *L. plantarum* (AMZ5) showed amylolytic potential through starch hydrolysis as it exhibited remarkable starch degradation capacity. [92].

Angiotensin-converting enzyme (ACE, EC 33.4.15.1, CD143) has a significant impact on the regulation of arterial blood pressure [93]. Inhibiting this enzyme can cause antihypertensive effects. Because of its role in the renin-angiotensin and kinin-nitric oxide systems, ACE-inhibitors are an ideal physiological target for clinical hypertension treatment [86]. However, ACE inhibitors that are currently available are synthetic pharmacological medicines that are not recommended for usage in healthy or low-risk populations due to side effects such as dry cough, skin rashes, and angioneurotic edema. As a result, producing safe and natural ACE inhibitors is critical for future hypertension therapy and prevention [94]. Previous studies show that ACE inhibitors are already been isolated from different products such as milk [95], cheese [96], yogurt [84], and other dairy products. The *L. helveticus* strains IMAU80872, IMAU80852, and IMAU80851 from fermented milk possessed a high ACE-inhibitory activity [86]. ACE inhibitory peptides were also isolated and identified from milk fermented with *L. delbrueckii* QS306 [85]. Moreover, ACE-inhibitory peptides account for the majority of bioactive peptides generated during yogurt fermentation processes. A strong link between *L. casei* LFTI® L26 growth and ACE inhibition in all yogurt samples was discovered during the initial stages of storage, compared to control yogurt, which reduced substantially after storage [84]. These previous researches prove that bioactive ACE-inhibitory producing *lactobacillus* strains have a great deal of potential for the improvement and production of functional dairy food products with antihypertensive effects.

The β-galactosidase enzyme, one of the glycosidases, is widely used in the dairy industry as well. These are produced by most *lactobacillus* species. Lactose, the primary carbohydrate in milk, is hydrolyzed by this enzyme into glucose and galactose, which may be absorbed via the intestinal epithelium. β-galactosidase involves two enzymatic activities: one hydrolyzes lactose and also cleaves cellobiose, cellotriose, cellotetrose, and to some extent cellulose, while the other splits β-glycosides [97]. High β-galactosidase activity observed in *L. rhamnosus* [88] and *L. bulgaricus* [83] has also been reported.

2.6 Exopolysaccharides (EPs)

Exopolysaccharides (EPs) are high–molecular, long-chain linear biopolymers containing side chains of homopolysaccharide or heteropolysaccharide carbohydrate units linked with α-glycosidic and β-glycosidic bonds [98]. The enzymes such as glycosyltransferase and glycoside hydrolase convert the sugar nucleotide precursors into EPs. EPs are "food-grade biopolymers," or extracellular biopolymers with a high molecular weight that are acquired from natural sources and produced during the metabolism of microorganisms [99].

Lactobacillus is one of the species of LAB that is frequently regarded as EPs-producing microorganisms (**Table 6**). An EPs termed as LPC-1 from *L. plantarum* C88 showed strong antioxidant activity and exhibited strong hydroxyl radical scavenging activity [107]. A novel EPs was also isolated from *L. plantarum* KX041 culture from a traditional Chinese pickle juice sample [109]. The EPs had a molecular weight of 38.67 KDa, which exhibited high thermal stability. EPS generated by *L. plantarum* has a good prospect to be utilized as natural antioxidants or functional additives in the food sector.

There has been a growing interest in using EPs-producing LAB for a variety of biological purposes. Among them, the anticancer action of EPs has attracted increasing attention. The EPs produced by *L. kefiri* MSR101 (MSR101 EPS) and its ability to inhibit the growth of HT-29 colon cancer cells were explored [103]. Structural analysis showed that MRS101 EPS is a heteropolysaccharide having a repeating unit of glucose and galactose and has a partial crystalline nature. *In-vitro* anticancer tests also showed significant anticancer action of MRS101 EPS against HT-29 cells. Moreover, a novel cell-bound exopolysaccharide (c-EPS) isolated from *L. helveticus* MB2–1 [102] showed high structural stability and may be used to make films and edible nanostructures for drug and food additive encapsulation. *In vitro* anticancer testing revealed that c-EPS exhibited substantial anticancer effects against human HepG-2 liver cancer, BGC-823 gastric cancer, and notably HT-29 colon cancer cells.

The utilization of probiotic microorganisms has been linked to a lower risk of cardiovascular disease, the leading cause of mortality and disability. The effect of dietary treatment of exopolysaccharide-producing probiotic *lactobacillus* on lipid metabolism and gut microbiota was investigated using apolipoprotein E (apoE)–deficient mice [104]. Dietary supplementation with a β-glucan–producing probiotic strain *L. mucosae* Dairy Product Culture Collection (DPC) 6426 resulted in lipid metabolism regulation in the mouse model of atherosclerosis. Several strains of *L. delbrueckii* subsp. *bulgaricus* isolated from homemade yogurt were also shown to produce EPs that can help cholesterol reduction [110]. The cholesterol removal mechanism, which involves binding or adhering to the bacterium cells, particularly to the EPS generated by the bacteria and enclosing the bacterial cells as a capsule, may be useful and relevant in human serum cholesterol management.

Lactobacillus strain	Exopolysaccharide	Reference
L. delbrueckii ssp. *bulgaricus* SRFM-1	Glucose and galactose	[100]
L. gasseri FR4	Glucose, mannose, galactose, rhamnose and fucose	[101]
L. helveticus MB2–1	Glucose, mannose, galactose, rhamnose, and arabinose (c-EPS)	[102]
L. kefiri MSR101	Galactose and glucose (MSR101)	[103]
L. mucosae (DPC) 6426	β-glucan	[104]
L. pentosus 14FE	Glucose	[105]
L. pentosus 68FE		
L. plantarum 47FE		
L. plantarum 301102S	Glucose and mannose	[106]
L. plantarum C88	Galactose and glucose (LPC-1)	[107]
L. plantarum H31	Glucose and mannose	[108]
L. plantarum KX041	Arabinose, mannose, glucose, and galactose	[109]

Table 6.
Example of exopolysaccharides produced by Lactobacillus *strains.*

2.7 Immune-modulating compounds

Different *lactobacillus* strains synthesize immune-modulating compounds that confer various health effects (**Table 7**). These most widely utilized probiotic agents promote intestinal microbiota and gut health and regulate the immune system in consumers. The immune system is modulated by probiotic bacteria, which control the synthesis of antibodies, interleukins, cytokines, and lymphocytes [121]. The probiotic bacteria interact with intestinal epithelial cells and generate immunomodulatory molecules, which activate the host immune response. By stimulating the production of interleukin-10 (IL-10) and immunoglobulin A antibodies (IgA), probiotics regulate immunity and inflammatory gene expression, reducing the host immunological response to infections [122]. IgA production, which is stimulated by dendritic cells, naive T cells, and B cells, promotes immune-modulatory effects as well and helps to eliminate pathogenic bacteria.

Numerous uropathogenic bacteria can interfere with the ability of the host to eliminate pathogens by subverting cellular functions. Probiotic *L. rhamnosus* GR-1 affected the immunological response of *E. coli* challenged of bladder cells by increasing NF-kappaB activation and TNF release [120]. The urogenital probiotic *L. rhamnosus* GR-1 regulated NF-kappaB activation by boosting TLR4 levels on bladder cells and modifying subsequent cytokine release from urothelial cells. These lactobacilli might help pathogen identification and infection control by affecting immunological factors like TLR4, which are crucial in the fight against infections.

Immune modulation and alterations in intestinal microbiota have been associated with probiotic administration, with implications for atopic dermatitis (AD). Oral administration of *L. paracasei* KBL382 was shown to significantly decrease AD-related skin lesions, epidermal thickening, immunoglobulin E levels in the blood, and immune cell infiltration [118]. Immunomodulatory activity in mice of *L. fermentum* JDFM216 was also shown to alter gut microbiota composition thus providing the advantage of improved health through better cognition, physiological behavior, and immunity [115].

Lactobacillus strain	Immune-modulating compound/mechanism	Reference
L. acidophilus DSM 32241	Increase in IgA and IgG levels	[111]
L. helveticus DSM 32242		
L. paracasei DSM 32243		
L. plantarum DSM 32244		
L. brevis DSM 27961		
L. brevis B13–2	Activated RAW 264.7 murine macrophages	[112]
L. fermentum BioE LF11	Inhibited secretion of lipopolysaccharide-induced pro-inflammatory cytokines IL-6 and TNF-α in RAW264.7 macrophages in vitro	[113]
L. plantarum BioE LPL59		
L. paracasei BioE LP08		
L. fermentum CECT5716	Modulation of intestinal cytokines IL-10 and IL-12	[114]
L. fermentum JDFM216	Increased sIgA	[115]
L. gasseri	Increased IFN-levels	[116]
L. fermentum		
L. plantarum		
L. paracasei subsp. *paracasei* G15	Lowered circulating LPS and inflammation cytokines, such as IL-1β and IL-8, and alleviated the inflammatory status and islet β-cell dysfunction	[117]
L. casei Q14		
L. paracasei KBL382	Increased IL-10 and growth factor-β	[118]
L. plantarum ST-III	Decrease the number of inflammatory cells	[119]
L. rhamnosus GR-1	TLR4	[120]

Table 7.
Immune-modulating compound/mechanism by Lactobacillus.

The probiotic potential of lactic acid bacteria strains isolated from Korean infant feces and Kimchi was also investigated [116]. The production of lymphocyte interferon (IFN) and cell proliferation were measured to assess the immunological modulatory activities of the strains. *L. gasseri, L. fermentum*, and *L. plantarum* strains all showed elevated IFN- levels and lymphocyte proliferation. In an *in vivo* model that assesses the impact of immune modifying lactobacilli on host life span using *Caenorhabditis elegans* as a model organism, feeding with *L. plantarum* CJLP133 and *L. fermentum* LA12 extended the average life span of the model host. Moreover, *L. brevis* B13–2 induced the expression of numerous cytokines (TNF-, IL-1, and IL-6) and iNOS [112].

2.8 Probiotic properties

Probiotics are live microorganisms that, when given in sufficient amounts, provide health benefits to the host. They create a favorable environment for the proper functioning of different metabolic activities in the gut, such as protein, carbohydrate, vitamin, and enzyme synthesis. Acids and the proteolytic activity of lactic acid bacteria inhibit harmful microorganisms in the intestine [123].

Colonized probiotic bacteria have a wide array of beneficial effects on the host cell, all of which are mediated via a large number of bioactive molecules. One of the mechanisms of probiotics includes competitive inhibition of the harmful bacteria by changing the pH and limiting the availability of oxygen, which leads to a less favorable environment in the intestine [123]. Probiotics also produce specific toxins with relatively narrow killing ranges, such as bacteriocins. It can also manufacture key micronutrients including vitamins, amino acids, and enzymes, boosting dietary nutrient bioavailability. Probiotics play an important role in stimulating the host immune system and enhancing the metabolic activity of carbohydrates as well [124].

Lactobacillus strains are essential components of the human and animal microbiome, and their varied impacts on host health have attracted a lot of interest [125]. *Lactobacillus* is one of the widely existing probiotic microorganisms. Probiotic lactobacilli have a significant and positive impact on the growth of the host, particularly in terms of enhancing body weight and size. The administration of probiotics *L. casei* variety *rhamnosus* to children with acute diarrhea showed a decrease in fecal lactoferrin and calprotectin concentrations and recovered faster as they exhibited significantly better appetite and oral intake, body weight gain, abdominal pain, bloating, and bowel movements [126]. An interleukin-22 (IL22)-secreting *L. reuteri* was shown to ameliorate non-alcoholic fatty liver disease [127].

2.9 Bio-converted metabolites

Dietary phytochemicals commonly occur in plant-based foods such as fruits and vegetables. These plant components with distinct bioactivities towards animal biochemistry and metabolism are being thoroughly investigated for their potential to deliver health advantages [128]. Phytochemicals often found as glyconjugates have lower bioactivity and bioavailability than their aglycone derivatives, which are smaller and less polar [129]. As a result, deglycosylation of plant glyconjugates (PGs) is recognized as a key factor in modulating their biological activity [130]. An *L. acidophilus* strain isolated from the human gut can activate dietary-relevant PG [131]. *L. acidophilus* was able to deglycosylate and externalize salicyl alcohol thus making it available for oxidation to salicylic acid by other microbial strains. This exhibits the ability of *Lactobacillus* to produce or mediate in the production of bio-converted metabolites (**Table 8**).

Resveratrol is a phytochemical found naturally in the grape skin and seeds, wine, berries, and medicinal plants [140]. It has antioxidant, anti-inflammatory, immunomodulatory, glycemic and lipid regulating, neuroprotective, and cardiovascular protective characteristics that can help protect against a wide range of chronic diseases [141]. The bioavailability and bioactivity of resveratrol are limited due to its presence in plants in glycosidic form as piceid. To get adequate amount and activity, deglycosylation of piceid to resveratrol from plant sources is necessary. A study by Basholli-Salihu et al. [132]investigated the enzymatic ability of probiotics to transform picied to resveratrol. Cell extracts of several probiotic strains from *Bifidobacterium* and *Lactobacillus* spp., including *B. infantis, Bifidobacterium bifidum, L. acidophilus, L. casei,* and *L. plantarum* have been shown to efficiently convert piceid to resveratrol. *L. acidophilus* effectively converts polydatin to resveratrol as well [131].

L. mucosae EPI2 was shown to convert daidzein to equol. Daidzein is a naturally occurring isoflavone but has lesser bioactivity than its deglycosylated form equol.

Lactobacillus strain	Bio-converted metabolites	Reference
L. acidophilus NCFM	Salicin	[131]
L. acidophilus	Resveratrol	[131]
L. acidophilus	Resveratrol	[132]
L. casei		
L. plantarum		
L. delbrueckii	β-maltooligosaccharides of glycitein and daidzein	[133]
L. intestinalis	Equol	[134]
L. kimchi JB301	Resveratrol	[135]
L. mucosae EPI2	Equol	[136]
L. plantarum	Daidzein	[137]
L. sakei subsp. sakei		
L. coryniformis		
L. plantarum K2–12	Daidzein and genistein	[138]
L. curvatus JD0–31		
Lactobacillus sp. Niu-O16	S-equol	[139]

Table 8.
Bio-converted metabolites produced by Lactobacillus.

Equol has significantly stronger estrogenic than daidzein [136]. Bovine rumen strain *Lactobacillus* sp. Niu-O16, in a mixed culture with human intestinal strain *Eggerthella* sp. Julong 732, has also been proven to successfully synthesize S-equol from daidzein through dihydrodaidzein under anaerobic conditions [139]. High amounts of equol have been shown to efficiently lower the risk of cancer.

3. Conclusion

This chapter summarizes the current knowledge of the existence of diverse bioactive molecules produced by the genus *lactobacillus*. The diverse bioactive compounds synthesized by *lactobacillus* confer health and nutritional benefits to humans and animals. These compounds include bacteriocin, bioactive peptides, short-chain fatty acids, vitamins, enzymes, exopolysaccharides immune-modulating compounds, and bio-converted molecules. Collectively, the physiological function and health of the consumers are enhanced and improved by these molecules.

Conflict of interest

The authors declare no conflict of interest.

Author details

Rodney H. Perez* and Amily E. Ancuelo
National Institute of Molecular Biology and Biotechnology (BIOTECH), University of the Philippines Los Baños (UPLB), Laguna, Philippines

*Address all correspondence to: rhperez@up.edu.ph

References

[1] Zheng J, Wittouck S, Salvetti E, Franz CMAP, Harris HMB, Mattarelli P, et al. A taxonomic note on the genus lactobacillus: Description of 23 novel genera, emended description of the genus lactobacillus Beijerinck 1901, and union of *Lactobacillaceae* and *Leuconostocaceae*. International Journal of Systematic and Evolutionary Microbiology. 2020;**70**:2782-2858. DOI: 10.1099/ijsem.0.004107

[2] EFSA Panel on Biological Hazards (BIOHAZ), Koutsoumanis K, Allende A, Alvarez-Ordóñez A, Bolton D, Bover-Cid S, et al. Update of the list of QPS-recommended biological agents intentionally added to food or feed as notified to EFSA 12: Suitability of taxonomic units notified to EFSA until march 2020. EFSA Journal. 2020;**18**:e06174. DOI: 10.2903/j.efsa.2020.6174

[3] Tamani RJ, Goh KKT, Brennan CS. Physico-chemical properties of sourdough bread production using selected lactobacilli starter cultures. Journal of Food Quality. 2013;**36**:245-252. DOI: 10.1111/jfq.12037

[4] Tachedijan G, Aldunate M, Bradshaw CS, Cone RA. The role of lactic acid production by probiotic lactobacillus species in vaginal health. Research in Microbiology. 2017;**168**:782-792. DOI: 10.1016/j.resmic.2017.04.001

[5] Plengvidhya V, Breidt F, Lu Z, Fleming HP. DNA fingerprinting of lactic acid bacteria in sauerkraut fermentations. Applied and Environmental Microbiology. 2007;**73**:7697-7702. DOI: 10.1128/AEM.01342-07

[6] Lonvaud-Funel A. Lactic acid bacteria in the quality improvement and depreciation of wine. Antonie van Leeuwenhoek International Journal of General and Molecular Microbiology. 1999;**76**:317-331. DOI: 10.1023/A:1002088931106

[7] Song Z, Du H, Zhang Y, Xu Y. Unraveling core functional microbiota in traditional solid-state fermentation by high-throughput amplicons and metatranscriptomics sequencing. Frontiers in Microbiology. 2017;**8**:1294. DOI: 10.3389/fmicb.2017.01294

[8] Shi Y, Singh A, Kitts D, Pratap-Singh A. Lactic acid fermentation: A novel approach to eliminate unpleasant aroma in pea protein isolates. LWT. 2021;**150**:111927. DOI: 10.1016/j.lwt.2021.111927

[9] Jia J, Ji B, Tian L, Li M, Lu M, Ding L, et al. Mechanism study on enhanced foaming properties of individual albumen proteins by lactobacillus fermentation. Food Hydrocolloids. 2021;**111**:106218. DOI: 10.1016/j.foodhyd.2020.106218

[10] Kemgang TS, Kapila S, Shanmugam VP, Kapila R. Cross-talk between probiotic lactobacilli and host immune system. Journal of Applied Microbiology. 2014;**117**:303-319. DOI: 10.1111/jam.12521

[11] Gao C, Major A, Rendon D, Lugo M, Jackson V, Shi Z, et al. Histamine H2 receptor-mediated suppression of intestinal inflammation by probiotic *Lactobacillus reuteri*. MBio. 2015;**6**:e01358-e01315. DOI: 10.1128/mBio.01358-15

[12] Delley M, Bruttin A, Richard M, Affolter M, Rezzonico E, Brück WM. *In vitro* activity of commercial probiotic

lactobacillus strains against uropathogenic *Escherichia coli*. FEMS Microbiology Letters. 2015;**362**:fnv096. DOI: 10.1093/femsle/fnv096

[13] Yue Y, Ye K, Lu J, Wang X, Zhang S, Liu L, et al. Probiotic strain *Lactobacillus plantarum* YYC-3 prevents colon cancer in mice by regulating the tumour microenvironment. Biomedicine & Pharmacotherapy. 2020;**127**:110159. DOI: 10.1016/j.biopha.2020.110159

[14] Barbosa MS, Todorov SD, Ivanova IV, Belguesmia Y, Choiset Y, Rabesona H, et al. Characterization of a two-peptide plantaricin produced by *Lactobacillus plantarum* MBSa4 isolated from Brazilian salami. Food Control. 2016;**60**:103-112. DOI: 10.1016/j.foodcont.2015.07.029

[15] Eid R, Jakee JE, Rashidy A, Asfour H, Omara S, Kandil MM, et al. Potential antimicrobial activities of probiotic lactobacillus strains isolated from raw milk. Journal of Probiotics & Health. 2016;**04**:138. DOI: 10.4172/2329-8901.1000138

[16] Neal-McKinney JM, Lu X, Duong T, Larson CL, Call DR, Shah DH, et al. Production of organic acids by probiotic lactobacilli can be used to reduce pathogen load in poultry. PLoS One. 2012;**7**:e43928. DOI: 10.1371/journal.pone.0043928

[17] Chen CC, Lai CC, Huang HL, Huang WY, Toh HS, Weng TC, et al. Antimicrobial activity of lactobacillus species against carbapenem-resistant *Enterobacteriaceae*. Frontiers in Microbiology. 2019;**10**:789. DOI: 10.3389/fmicb.2019.00789

[18] Perez RH, Zendo T, Sonomoto K. Novel bacteriocins from lactic acid bacteria (LAB): Various structures and applications. Microbial Cell Factories. 2014;**13**:S3. DOI: 10.1186/1475-2859-13-S1-S3

[19] Gálvez A, Abriouel H, López RL, Ben ON. Bacteriocin-based strategies for food biopreservation. International Journal of Food Microbiology. 2007;**120**:51-70. DOI: 10.1016/j.ijfoodmicro.2007.06.001

[20] Cotter P, Hill C, Ross R. Bacteriocins: Developing innate immunity for foods. Nature Reviews. Microbiology. 2005;**3**:777-788. DOI: 10.1038/nrmicro1273

[21] Hussein WE, Huang E, Ozturk I, Somogyi Á, Yang X, Liu B, et al. Genome-guided mass spectrometry expedited the discovery of paraplantaricin TC318, a lantibiotic produced by *Lactobacillus paraplantarum* strain isolated from cheese. Frontiers in Microbiology. 2020;**11**:1381. DOI: 10.3389/fmicb.2020.01381

[22] Flórez AB, Mayo B. Genome analysis of *Lactobacillus plantarum* LL441 and genetic characterisation of the locus for the lantibiotic plantaricin C. Frontiers in Microbiology. 2018;**9**:1916. DOI: 10.3389/fmicb.2018.01916

[23] Mortvedt-Abildgaard CI, Nissen-Meyer J, Jelle B, Grenov B, Skaugen M, Nes IF. Production and pH-dependent bactericidal activity of lactocin S, a lantibiotic from *Lactobacillus sake* L45. Applied and Environmental Microbiology. 1995;**61**:175-179. DOI: 10.1128/aem.61.1.175-179.1995

[24] Vermeulen R, Deane S, Dicks L, Rohwer J, van Staden ADP. Manganese privation-induced transcriptional upregulation of the class IIa bacteriocin plantaricin 423 in *Lactobacillus plantarum* strain 423. Applied and Environmental Microbiology. 2021;**87**:e0097621. DOI: 10.1128/aem.00976-21

[25] Wang Y, Qin Y, Xie Q, Zhang Y, Hu J, Li P. Purification and characterization

of plantaricin LPL-1, a novel class IIa bacteriocin produced by *Lactobacillus plantarum* LPL-1 isolated from fermented fish. Frontiers in Microbiology. 2018;**9**:2276. DOI: 10.3389/fmicb.2018.02276

[26] Jeong YJ, Moon GS. Antilisterial bacteriocin from *Lactobacillus rhamnosus* CJNU 0519 presenting a narrow antimicrobial spectrum. Korean Journal for Food Science of Animal Resources. 2015;**35**:137-142. DOI: 10.5851/kosfa.2015.35.1.137

[27] Kasuga G, Tanaka M, Harada Y, Nagashima H, Yamato T, Wakimoto A, et al. Homologous expression and characterization of gassericin T and gassericin S, a novel class IIb bacteriocin produced by *Lactobacillus gasseri* LA327. Applied and Environmental Microbiology. 2019;**85**:e02815-e02818. DOI: 10.1128/AEM.02815-18

[28] Garcia-Gutierrez E, O'Connor PM, Colquhoun IJ, Vior NM, Rodríguez JM, Mayer MJ, et al. Production of multiple bacteriocins, including the novel bacteriocin gassericin M, by *Lactobacillus gasseri* LM19, a strain isolated from human milk. Applied Microbiology and Biotechnology. 2020;**104**:3869-3884. DOI: 10.1007/s00253-020-10493-3

[29] Acedo JZ, van Belkum MJ, Lohans CT, McKay RT, Miskolzie M, Vederas JC. Solution structure of acidocin B, a circular bacteriocin produced by *Lactobacillus acidophilus* M46. Applied and Environmental Microbiology. 2015;**81**:2910-2918. DOI: 10.1128/AEM.04265-14

[30] Borrero J, Kelly E, O'Connor PM, Kelleher P, Scully C, Cotter PD, et al. Plantaricyclin A, a novel circular bacteriocin produced by *Lactobacillus plantarum* NI326: Purification, characterization, and heterologous production. Applied and Environmental Microbiology. 2018;**84**:e01801-e01817. DOI: 10.1128/AEM.01801-17

[31] Chee Gor M, Golneshin A, Van TTH, Moore RJ, Smith AT. Cloning and functional expression of a food-grade circular bacteriocin, plantacyclin B21AG, in probiotic *Lactobacillus plantarum* WCFS1. PLoS One. 2020;**15**:e0232806. DOI: 10.1371/journal.pone.0232806

[32] Sawa N, Koga S, Okamura K, Ishibashi N, Zendo T, Sonomoto K. Identification and characterization of novel multiple bacteriocins produced by *Lactobacillus sakei* D98. Journal of Applied Microbiology. 2013;**115**:61-69. DOI: 10.1111/jam.12226

[33] O'Connor PM, O'Shea EF, Cotter PD, Hill C, Ross RP. The potency of the broad spectrum bacteriocin, bactofencin A, against *staphylococci* is highly dependent on primary structure, N-terminal charge and disulphide formation. Scientific Reports. 2018;**8**:11833. DOI: 10.1038/s41598-018-30271-6

[34] Sun Z, Wang X, Zhang X, Wu H, Zou Y, Li P, et al. Class III bacteriocin Helveticin-M causes sublethal damage on target cells through impairment of cell wall and membrane. Journal of Industrial Microbiology & Biotechnology. 2018;**45**:213-227. DOI: 10.1007/s10295-018-2008-6

[35] Rushdy AA, Gomaa EZ. Antimicrobial compounds produced by probiotic *Lactobacillus brevis* isolated from dairy products. Annales de Microbiologie. 2013;**63**:81-90. DOI: 10.1007/s13213-012-0447-2

[36] Vaughan A, Eijsink VGH, Sinderen DV. Functional characterization of a composite bacteriocin locus from malt isolate *Lactobacillus sakei* 5. Applied

and Environmental Microbiology. 2003;**69**:7194-7720. DOI: 10.1128/AEM.69.12.7194-7203.2003

[37] Le NTT, Bach LG, Nguyen DC, Le THX, Pham KH, Nguyen DH, et al. Evaluation of factors affecting antimicrobial activity of bacteriocin from *Lactobacillus plantarum* microencapsulated in alginate-gelatin capsules and its application on pork meat as a bio-preservative. International Journal of Environmental Research and Public Health. 2019;**16**:1017. DOI: 10.3390/ijerph16061017

[38] Todorov SD. Bacteriocin production by *Lactobacillus plantarum* AMA-K isolated from Amasi, a Zimbabwean fermented milk product and study of the adsorption of bacteriocin AMA-K to Listeria sp. Brazilian Journal of Microbiology. 2008;**39**:178-187. DOI: 10.1590/S1517-83822008000100035

[39] Milioni C, Martínez B, Degl'Innocenti S, Turchi B, Fratini F, Cerri D, et al. A novel bacteriocin produced by *Lactobacillus plantarum* LpU4 as a valuable candidate for biopreservation in artisanal raw milk cheese. Dairy Science & Technology. 2015;**95**:479-494. DOI: 10.1007/s13594-015-0230-9

[40] Zangeneh M, Khorrami S, Khaleghi M. Bacteriostatic activity and partial characterization of the bacteriocin produced by *L. plantarum* sp. isolated from traditional sourdough. Food Science & Nutrition. 2020;**8**:6023-6030. DOI: 10.1002/fsn3.1890

[41] Ravel J, Gajer P, Abdo Z, Schneider GM, Koenig SS, McCulle SL, et al. Vaginal microbiome of reproductive-age women. Proceedings of the National Academy of Sciences of the United States of America. 2011;**108**:4680-4687. DOI: 10.1073/pnas.1002611107

[42] Dong Q, Nelson DE, Toh E, Diao L, Gao X, Fortenberry JD, et al. The microbial communities in male first catch urine are highly similar to those in paired urethral swab specimens. PLoS One. 2011;**6**:e19709. DOI: 10.1371/journal.pone.0019709

[43] Gupta K, Stapleton AE, Hooton TM, Roberts PL, Fennell CL, Stamm WE. Inverse association of H2O2-producing lactobacilli and vaginal *Escherichia coli* colonization in women with recurrent urinary tract infections. The Journal of Infectious Diseases. 1998;**178**:446-450. DOI: 10.1086/515635

[44] Gaspar C, Donders GG, Palmeira-de-Oliveira R, Queiroz JA, Tomaz C, Martinez-de-Oliveira J, et al. Bacteriocin production of the probiotic *Lactobacillus acidophilus* KS400. AMB Express. 2018;**8**:153. DOI: 10.1186/s13568-018-0679-z

[45] Pessione E. Lactic acid bacteria contribution to gut microbiota complexity: Lights and shadows. Frontiers in Cellular and Infection Microbiology. 2012;**2**:86. DOI: 10.3389/fcimb.2012.00086

[46] Pessione E, Cirrincione S. Bioactive molecules released in food by lactic acid bacteria: Encrypted peptides and biogenic amines. Frontiers in Microbiology. 2016;**7**:876. DOI: 10.3389/fmicb.2016.00876

[47] Fernández M, Zúñiga M. Amino acid catabolic pathways of lactic acid bacteria. Critical Reviews in Microbiology. 2006;**32**(3):155-183. DOI: 10.1080/10408410600880643

[48] Hayes M, Ross RP, Fitzgerald GF, Stanton C. Putting microbes to work: Dairy fermentation, cell factories and bioactive peptides. Part I: overview. Biotechnology Journal. 2007;**2**:426-434. DOI: 10.1002/biot.200600246

[49] Raveschot C, Cudennec B, Coutte F, Flahaut C, Fremont M, Drider D, et al. Production of bioactive peptides by lactobacillus *species*: From gene to application. Frontiers in Microbiology. 2018;**9**:2354. DOI: 10.3389/fmicb.2018.02354

[50] Yamamoto N. Antihypertensive peptides derived from food proteins. Biopolymers. 1997;**43**:129-134. DOI: 10.1002/(SICI)1097-0282(1997)43:2<129::AID-BIP5>3.0.CO;2-X

[51] Chandra P, Vij S. Molecular characterization and identification of bioactive peptides producing *Lactobacillus* sps. Based on 16S rRNA gene sequencing. Food Biotechnology. 2018;**32**:1-14. DOI: 10.1080/08905436.2017.1413657

[52] Skrzypczak K, Gustaw W, Szwajgier D, Fornal E, Waśko A. κ-Casein as a source of short-chain bioactive peptides generated by *Lactobacillus helveticus*. Journal of Food Science and Technology. 2017;**54**:3679-3688. DOI: 10.1007/s13197-017-2830-2

[53] Raveschot C, Deracinois B, Bertrand E, Flahaut C, Frémont M, Drider D, et al. Integrated continuous bioprocess development for ace-inhibitory peptide production by *Lactobacillus helveticus* strains in membrane bioreactor. Frontiers in Bioengineering and Biotechnology. 2020;**8**:585815. DOI: 10.3389/fbioe.2020.585815

[54] Adams C, Sawh F, Green-Johnson JM, Jones Taggart H, Strap JL. Characterization of casein-derived peptide bioactivity: Differential effects on angiotensin-converting enzyme inhibition and cytokine and nitric oxide production. Journal of Dairy Science. 2020;**103**:5805-5815. DOI: 10.3168/jds.2019-17976

[55] Rokka T, Syväoja EL, Tuominen J, Korhonen H. Release of bioactive peptides by enzymatic proteolysis of *Lactobacillus* GG fermented UHT milk. Milchwissenschaft. 1997;**52**:675-678

[56] Aguilar-Toalá JE, Santiago-LópezL, PeresC, PeresC, GarciaH, Vallejo-Cordoba B, et al. Assessment of multifunctional activity of bioactive peptides derived from fermented milk by specific *Lactobacillus plantarum* strains. Journal of Dairy Science. 2016;**100**:65-75. DOI: 10.3168/jds.2016-11846

[57] Singh BP, Vij S. Growth and bioactive peptides production potential of *Lactobacillus plantarum strain* C2 in soy milk: A LC-MS/MS based revelation for peptides biofunctionality. LWT- Food Science and Technology. 2017;**86**:293-301. DOI: 10.1016/j.lwt.2017.08.013

[58] Galli V, Mazzoli L, Luti S, Venturi M, Guerrini S, Paoli P, et al. Effect of selected strains of lactobacilli on the antioxidant and anti-inflammatory properties of sourdough. International Journal of Food Microbiology. 2018;**286**:55-65. DOI: 10.1016/j.ijfoodmicro.2018.07.018

[59] Clare DA, Swaisgood HE. Bioactive milk peptides: A prospectus. Journal of Dairy Science. 2000;**83**:1187-1195. DOI: 10.3168/jds.S0022-0302(00)74983-6

[60] Fan M, Guo T, Li W, Chen J, Li F, Wang C, et al. Isolation and identification of novel casein-derived bioactive peptides and potential functions in fermented casein with *Lactobacillus helveticus*. Food Science and Human Wellness. 2019;**8**:156-176. DOI: 10.1016/j.fshw.2019.03.010

[61] Nakamura Y, Yamamoto N, Sakai K, Okubo A, Yamazaki S, Takano T. Purification and characterization of

angiotensin I-converting enzyme inhibitors from sour milk. Journal of Dairy Science. 1995;**78**:777-783. DOI: 10.3168/jds.S0022-0302(95)76689-9

[62] Gobbetti M, Ferranti P, Smacchi E, Goffredi F, Addeo F. Production of angiotensin-I-converting-enzyme-inhibitory peptides in fermented milks started by *Lactobacillus delbrueckii* subsp. *bulgaricus* SS1 and *Lactococcus lactis* subsp. cremoris FT4. Applied and Environmental Microbiology. 2000;**66**:3898-3904. DOI: 10.1128/AEM.66.9.3898-3904.2000

[63] Topping DL, Clifton PM. Short-chain fatty acids and human colonic function: Roles of resistant starch and nonstarch polysaccharides. Physiological Reviews. 2001;**81**:1031-1064. DOI: 10.1152/physrev.2001.81.3.1031

[64] Qian Z, Zhu H, Zhao D, Ping Y, Gao F, Lu C, et al. Probiotic *Lactobacillus* sp. strains inhibit growth, adhesion, biofilm formation, and gene expression of bacterial vaginosis-inducing *Gardnerella vaginalis*. Microorganisms. 2021;**9**:728. DOI: 10.3390/microorganisms9040728

[65] Cremon C, Guglielmetti S, Gargari G, Taverniti V, Castellazzi AM, Valsecchi C, et al. Effect of *Lactobacillus paracasei* CNCM I-1572 on symptoms, gut microbiota, short chain fatty acids, and immune activation in patients with irritable bowel syndrome: A pilot randomized clinical trial. United European Gastroenterology Journal. 2018;**6**:604-613. DOI: 10.1177/2050640617736478

[66] Kao L, Liu TH, Tsai TY, Pan TM. Beneficial effects of the commercial lactic acid bacteria product, Vigiis 101, on gastric mucosa and intestinal bacterial flora in rats. Journal of Microbiology, Immunology, and Infection. 2020;**53**:266-273. DOI: 10.1016/j.jmii.2018.06.002

[67] Jang HJ, Lee NK, Paik HD. *Lactobacillus plantarum* G72 showing production of folate and short-chain fatty acids. Microbiology and Biotechnology Letters. 2021;**49**:18-23. DOI: 10.48022/mbl.2009.09010

[68] Zhang C, Zhang Y, Liu G, Li W, Xia S, Li H, et al. Effects of soybean protein isolates and peptides on the growth and metabolism of *Lactobacillus rhamnosus*. Journal of Functional Foods. 2021;**77**:104335. DOI: 10.1016/j.jff.2020.104335

[69] Malhotra MKI. Screening and *in-vitro* analysis of *Lactobacillus reuteri* strains for short chain fatty acids production, stability and therapeutic potentials in colorectal cancer. Journal of Bioequivalence & Bioavailability. 2015;**07**:01. DOI: 10.4172/jbb.1000212

[70] Meimandipour A, Shuhaimi M, Soleimani AF, Azhar K, Hair-Bejo M, Kabeir BM, et al. Selected microbial groups and short-chain fatty acids profile in a simulated chicken cecum supplemented with two strains of lactobacillus. Poultry Science. 2010;**89**:470-476. DOI: 10.3382/ps.2009-00495

[71] Wang G, Zhu G, Chen C, Zheng Y, Ma F, Zhao J, et al. Lactobacillus strains derived from human gut ameliorate metabolic disorders via modulation of gut microbiota composition and short-chain fatty acids metabolism. Beneficial Microbes. 2021;**12**:267-281. DOI: 10.3920/BM2020.0148

[72] Hati S, Patel M, Mishra B, Das S. Short-chain fatty acid and vitamin production potentials of lactobacillus isolated from fermented foods of Khasi tribes, Meghalaya, India. Annales de

Microbiologie. 2019;**69**:1191-1199. DOI: 10.1007/s13213-019-01500-8

[73] Hamzehlou P, Sepahy AA, Mehrabian S, Hosseini F. Production of vitamins B3, B6 and B9 by lactobacillus isolated from traditional yogurt samples from 3 cities in Iran, winter 2016. Applied Food Biotechnology. 2018;**5**: 105-118. DOI: 10.22037/afb.v%vi%i.18651

[74] Levit R, Savoy de Giori G, de Moreno de LeBlanc A, LeBlanc JG. Effect of riboflavin-producing bacteria against chemically induced colitis in mice. Journal of Applied Microbiology. 2018;**124**:232-240. DOI: 10.1111/jam.13622

[75] Li P, Gu Q, Yang L, Yu Y, Wang Y. Characterization of extracellular vitamin B12 producing *Lactobacillus plantarum* strains and assessment of the probiotic potentials. Food Chemistry. 2017;**234**:494-501. DOI: 10.1016/j.foodchem.2017.05.037

[76] Bhushan B, Tomar SK, Chauhan A. Techno-functional differentiation of two vitamin B12 producing *Lactobacillus plantarum* strains: An elucidation for diverse future use. Applied Microbiology and Biotechnology. 2017;**101**:697-709. DOI: 10.1007/s00253-016-7903-z

[77] Gu Q, Zhang C, Song D, Li P, Zhu X. Enhancing vitamin B12 content in soy-yogurt by *Lactobacillus reuteri*. International Journal of Food Microbiology. 2015;**206**:56-59. DOI: 10.1016/j.ijfoodmicro.2015.04.033

[78] De Angelis M, Bottacini F, Fosso B, Kelleher P, Calasso M, Di Cagno R, et al. *Lactobacillus rossiae*, a vitamin B12 producer, represents a metabolically versatile species within the genus lactobacillus. PLoS One. 2014;**9**:e107232. DOI: 10.1371/journal.pone.0107232

[79] LeBlanc JG, Levit R, Savoy de Giori G, de Moreno de LeBlanc A. Application of vitamin-producing lactic acid bacteria to treat intestinal inflammatory diseases. Applied Microbiology and Biotechnology 2020;104:3331-3337. DOI: 10.1007/s00253-020-10487-1

[80] Maske BL, de Melo Pereira GV, Vale A d S, de Carvalho Neto DP, Karp SG, Viesser JA, et al. A review on enzyme-producing lactobacilli associated with the human digestive process: From metabolism to application. Enzyme and Microbial Technology. 2021;**149**:109836. DOI: 10.1016/j.enzmictec.2021.109836

[81] Akin-Osanaiye BC, Azeez BT, Olobayotan IW. Evaluation of invertase and amylase activities of lactic acid bacteria isolated from 'pupuru' (an indigenous African fermented cassava staple food). Asian Journal of Biochemistry. 2019;**5**:1-8. DOI: 10.9734/ajrb/2019/v5i330090

[82] Sieo CC, Abdullah N, Tan WS, Ho YW. *In vivo* study on the persistence of transformed β-glucanase-producing lactobacillus strains in the gastrointestinal tract of chickens. Journal of Animal and Feed Sciences. 2006;**15**:261-274. DOI: 10.22358/jafs/66898/2006

[83] Zhao R, Duan F, Yang J, Xiao M, Lu L. Integrated utilization of dairy whey in probiotic β-galactosidase production and enzymatic synthesis of galacto-oligosaccharides. Catalysts. 2020;**11**:658. DOI: 10.3390/catal11060658

[84] Rezaei A, Amirdivani S, Khosrowshahi Asl A, Malekinejad H, Zomorodi S, Hosseinmardi F. Inhibition of the angiotensin I converting enzyme (ACE) and proteolysis of non-fat probiotic yogurt. Brazilian Journal of Food Technology. 2019;**22**:e2018234. DOI: 10.1590/1981-6723.23418

[85] Wu N, Xu W, Liu K, Xia Y, Shuangquan. Angiotensin-converting enzyme inhibitory peptides from *Lactobacillus delbrueckii* QS306 fermented milk. Journal of Dairy Science. 2019;**102**:5913-5921. DOI: 10.3168/jds.2018-15901

[86] Chen Y, Li C, Xue J, Kwok LY, Yang J, Zhang H, et al. Characterization of angiotensin-converting enzyme inhibitory activity of fermented milk produced by *Lactobacillus helveticus*. Journal of Dairy Science. 2015;**98**:5113-5124. DOI: 10.3168/jds.2015-9382

[87] Singh B, Katiyar D, Chauhan RS, Kharwar RK, Lall AM. Influence of UV treatment on α-galactosidase produced by *Lactobacillus plantarum*. Journal of Pure and Applied Microbiology. 2013;7:595-601

[88] Lima M, Souza K, Pastrana L, Vieira Soares MT, Porto A. *In vitro* digestion as a tool for functional isolation of a probiotic potential *Lactobacillus rhamnosus*. Research, Society and Development. 2020;**9**:e3119108544. DOI: 10.33448/rsd-v9i10.8544

[89] Padmavathi T, Bhargavi R, Priyanka PR, Niranjan NR, Pavitra PV. Screening of potential probiotic lactic acid bacteria and production of amylase and its partial purification. Journal, Genetic Engineering & Biotechnology. 2018;**16**:357-362. DOI: 10.1016/j.jgeb.2018.03.005

[90] de Souza PM, de Oliveira MP. Application of microbial α-amylase in industry - A review. Brazilian Journal of Microbiology. 2010;**41**:850-861. DOI: 10.1590/S1517-83822010000400004

[91] Singh SK, Ahmed SU, Pandey A. Metabolic engineering approaches for lactic acid production. Process Biochemistry. 2006;**41**:991-1000. DOI: 10.1016/j.procbio.2005.12.004

[92] Amapu T, Baba A, Ado S, Abdullahi I, Hycinth D. Amylolytic potential of lactic acid bacteria isolated from wet milled cereals, cassava flour and fruits. British Microbiology Research Journal. 2016;**13**:1-8. DOI: 10.9734/BMRJ/2016/24509

[93] Tikhomirova VE, Kost OA, Kryukova OV, Golukhova EZ, Bulaeva NI, Zholbaeva AZ, et al. ACE phenotyping in human heart. PLoS One. 2017;**12**:e0181976. DOI: 10.1371/journal.pone.0181976

[94] Jao C, Huang S, Hsu K. Angiotensin I-converting enzyme inhibitory peptides: Inhibition mode, bioavailability, and antihypertensive effects. Biomedicine. 2012;**2**:130-136. DOI: 10.1016/j.biomed.2012.06.005

[95] Wang J, Li C, Xue J, Yang J, Zhang Q, Zhang H, et al. Fermentation characteristics and angiotensin I-converting enzyme-inhibitory activity of *Lactobacillus helveticus* isolate H9 in cow milk, soy milk, and mare milk. Journal of Dairy Science. 2015;**98**:3655-3664. DOI: 10.3168/jds.2015-9336

[96] Yousefi L, Habibi Najafi M, Edalatian Dovom M, Mortazavian A. Production of angiotensin-converting enzyme inhibitory peptides in Iranian ultrafiltered white cheese prepared with *Lactobacillus brevis* KX572382. International Journal of Food Science. 2020;**56**:2530-2538. DOI: 10.1111/ijfs.14891

[97] Troelsen JT. Adult-type hypolactasia and regulation of lactase expression. Biochimica et Biophysica Acta. 2005;**1723**:19-32. DOI: 10.1016/j.bbagen.2005.02.003

[98] Berthold-Pluta A, St. Pluta A, Garbowska M, Stasiak L. Exopolysaccharide-producing lactic acid bacteria - health-promoting properties

and application in the dairy industry. Advances in Microbiology. 2019;**58**:191-204. DOI: 10.21307/PM-2019.58.2.191

[99] Zhou K, Zeng Y, Yang M, Chen S, He L, Ao X, et al. Production, purification and structural study of an exopolysaccharide from *Lactobacillus plantarum* BC-25. Carbohydrate Polymers. 2016;**144**:205-214. DOI: 10.1016/j.carbpol.2016.02.067

[100] Tang W, Han S, Zhou J, Xu Q, Dong M, Fan X, et al. Selective fermentation of *Lactobacillus delbrueckii* ssp. *Bulgaricus* SRFM-1 derived exopolysaccharide by lactobacillus and *Streptococcus* strains revealed prebiotic properties. Journal of Functional Foods. 2020;**69**:103952. DOI: 10.1016/j.jff.2020.103952

[101] Rani RP, Anandharaj M, David RA. Characterization of a novel exopolysaccharide produced by *Lactobacillus gasseri* FR4 and demonstration of its in vitro biological properties. International Journal of Biological Macromolecules. 2018;**109**:772-783. DOI: 10.1016/j.ijbiomac.2017.11.062

[102] Li W, Xia X, Tang W, Ji J, Rui X, Chen X, et al. Structural characterization and anticancer activity of cell-bound exopolysaccharide from *Lactobacillus helveticus* MB2-1. Journal of Agricultural and Food Chemistry. 2015;**63**:3454-3463. DOI: 10.1021/acs.jafc.5b01086

[103] Riaz MSR, Mehwish HM, Fang H, Padhiar AA, Zend X, Khurshid M, et al. Characterization and anti-tumor activity of exopolysaccharide produced by *Lactobacillus kefiri* isolated from Chinese kefir grains. Journal of Functional Foods. 2019;**63**:10588. DOI: 10.1016/j.jff.2019.103588

[104] London LL, Kumar AHS, Wall R, Casey PG, O'Sullivan O, Shanahan F, et al. Exopolysaccharide-producing probiotic lactobacilli reduce serum cholesterol and modify enteric microbiota in ApoE-deficient mice. The Journal of Nutrition. 2014;**144**:1956-1962. DOI: 10.3945/jn.114.191627

[105] Abo Saif FAA, Sakr EAE. Characterization and bioactivities of exopolysaccharide produced from probiotic *Lactobacillus plantarum* 47FE and *Lactobacillus pentosus* 68FE. Bioactive Carbohydrates and Dietary Fibre. 2020;**24**:100231. DOI: 10.1016/j.bcdf.2020.100231

[106] Tsuda H, Miyamoto T. Production of exopolysaccharide by *Lactobacillus plantarum* and the prebiotic activity of the exopolysaccharide. Food Science and Technology Research. 2010;**16**:87-92. DOI: 10.3136/fstr.16.87

[107] Zhang L, Liu C, Li D, Zhao Y, Zhang X, Zeng X, et al. Antioxidant activity of an exopolysaccharide isolated from *Lactobacillus plantarum* C88. International Journal of Biological Macromolecules. 2013;**54**:270-275. DOI: 10.1016/j.ijbiomac.2012.12.037

[108] Huang Z, Lin F, Zhu X, Zhang C, Jiang M, Lu Z. An exopolysaccharide from *Lactobacillus plantarum* H31 in pickled cabbage inhibits pancreas α-amylase and regulating metabolic markers in HepG2 cells by AMPK/PI3K/Akt pathway. International Journal of Biological Macromolecules. 2020;**143**:775-784. DOI: 10.1016/j.ijbiomac.2019.09.137

[109] Wang X, Shao C, Liu L, Guo X, Xu Y, Lü X. Optimization, partial characterization and antioxidant activity of an exopolysaccharide from *Lactobacillus plantarum* KX041. International Journal of Biological Macromolecules. 2017;**103**:1173-1184. DOI: 10.1016/j.ijbiomac.2017.05.118

[110] Tok E, Aslim B. Cholesterol removal by some lactic acid bacteria that can be used as probiotic. Microbiology and Immunology. 2010;**54**:257-264. DOI: 10.1111/j.1348-0421.2010.00219.x

[111] Rossi G, Pengo G, Galosi L, Berardi S, Tambella AM, Attili AR, et al. Effects of the probiotic mixture Slab51® (SivoMixx®) as food supplement in healthy dogs: Evaluation of fecal microbiota, clinical parameters and immune function. Frontiers in Veterinary Science. 2020;7:613. DOI: 10.3389/fvets.2020.00613

[112] Song MW, Chung Y, Kim K, Hong WS, Chang HJ, Paik H. Probiotic characteristics of *Lactobacillus brevis* B13-2 isolated from kimchi and investigation of antioxidant and immune-modulating abilities of its heat-killed cells. LWT. 2020;**128**:109452. DOI: 10.1016/j.lwt.2020.109452

[113] Kook SY, Chung EC, Lee Y, Lee DW, Kim S. Isolation and characterization of five novel probiotic strains from Korean infant and children faeces. PLoS One. 2019;**14**:e0223913. DOI: 10.1371/journal.pone.0223913

[114] Azagra-Boronat I, Tres A, Massot-Cladera M, Franch À, Castell M, Guardiola F, et al. *Lactobacillus fermentum* CECT5716 supplementation in rats during pregnancy and lactation impacts maternal and offspring lipid profile, immune system and microbiota. Cell. 2020;**9**:575. DOI: 10.3390/cells9030575

[115] Park MR, Shin M, Mun D, Jeong SY, Jeong DY, Song M, et al. Probiotic *Lactobacillus fermentum* strain JDFM216 improves cognitive behavior and modulates immune response with gut microbiota. Scientific Reports. 2020;**10**:21701. DOI: 10.1038/s41598-020-77587-w

[116] Lee J, Yun HS, Cho KW, Oh S, Kim SH, Chun T, et al. Evaluation of probiotic characteristics of newly isolated *Lactobacillus* spp.: Immune modulation and longevity. International Journal of Food Microbiology. 2011;**148**:80-86. DOI: 10.1016/j.ijfoodmicro.2011.05.003

[117] Tian P, Li B, He C, Song W, Hou A, Tian S, et al. Antidiabetic (type 2) effects of Lactobacillus G15 and Q14 in rats through regulation of intestinal permeability and microbiota. Food & Function. 2016;7:3789-3797. DOI: 10.1039/c6fo00831c

[118] Kim WK, Jang YJ, Han DH, Jeon K, Lee C, Han HS, et al. *Lactobacillus paracasei* KBL382 administration attenuates atopic dermatitis by modulating immune response and gut microbiota. Gut Microbes. 2020;**12**:1-14. DOI: 10.1080/19490976.2020.1819156

[119] Zang L, Ma Y, Huang W, Ling Y, Sun L, Wang X, et al. Dietary lactobacillus *plantarum* ST-III alleviates the toxic effects of triclosan on zebrafish (Danio rerio) via gut microbiota modulation. Fish & Shellfish Immunology. 2019;**84**:1157-1169. DOI: 10.1016/j.fsi.2018.11.007

[120] Karlsson M, Scherbak N, Reid G, Jass J. *Lactobacillus rhamnosus* GR-1 enhances NF-kappaB activation in *Escherichia coli*-stimulated urinary bladder cells through TLR4. BMC Microbiology. 2012;**12**:15. DOI: 10.1186/1471-2180-12-15

[121] Nagpal R, Kumar A, Kumar M, Behare PV, Jain S, Yadav H. Probiotics, their health benefits and applications for developing healthier foods: A review. FEMS Microbiology Letters. 2012;**334**:1-15. DOI: 10.1111/j.1574-6968.2012.02593.x

[122] Kawashima T, Ikari N, Kouchi T, Kowatari Y, Kubota Y, Shimojo N, et al. The molecular mechanism for activating IgA production by *Pediococcus acidilactici* K15 and the clinical impact in a randomized trial. Scientific Reports. 2018;**8**:5065. DOI: 10.1038/s41598-018-23404-4

[123] Schepper JD, Irwin R, Kang J, Dagenais K, Lemon T, Shinouskis A, et al. Probiotics in gut-bone signaling. Advances in Experimental Medicine and Biology. 2017;**1033**:225-247. DOI: 10.1007/978-3-319-66653-2_11

[124] George Kerry R, Patra JK, Gouda S, Park Y, Shin HS, Das G. Benefaction of probiotics for human health: A review. Journal of Food and Drug Analysis. 2018;**26**:927-939. DOI: 10.1016/j.jfda.2018.01.002

[125] Zhang Z, Lv J, Pan L, Zhang Y. Roles and applications of probiotic lactobacillus strains. Applied Microbiology and Biotechnology. 2018;**102**:8135-8143. DOI: 10.1007/s00253-018-9217-9

[126] Lai H, Chiu C, Kong M, Chang C, Chen C. Probiotic *Lactobacillus casei*: Effective for managing childhood diarrhea by altering gut microbiota and attenuating fecal inflammatory markers. Nutrients. 2019;**11**:1150. DOI: 10.3390/nu11051150

[127] Oh J, Schueler K, Stapleton D, Alexander L, Yen C, Keller MP, et al. Secretion of recombinant Interleukin-22 by engineered *Lactobacillus reuteri* reduces fatty liver disease in a mouse model of diet-induced obesity. ASM Journals. 2020;**5**:e00183-e00120. DOI: 10.1128/msphere.00183-20

[128] Dillard CJ, German JB. Phytochemicals: Nutraceuticals and human health. Journal of the Science of Food and Agriculture. 2000;**80**:1744-1756. DOI: 10.1002/1097-0010(20000915)80:12<1744::AID-JSFA725>3.0.CO;2-W

[129] Landete JM, Curiel JA, Rodríguez H, de las Rivas B, Muñoz R. Aryl glycosidases from *Lactobacillus plantarum* increase antioxidant activity of phenolic compounds. Journal of Functional Foods. 2014;7:322-329. DOI: 10.1016/j.jff.2014.01.028

[130] Possemiers S, Rabot S, Espín JC, Bruneau A, Philippe C, González-Sarrías A, et al. Eubacterium limosum activates isoxanthohumol from hops (*Humulus lupulus* L.) into the potent phytoestrogen 8-prenylnaringenin in vitro and in rat intestine. The Journal of Nutrition. 2008;**138**:1310-1316. DOI: 10.1093/jn/138.7.1310

[131] Theilmann MC, Goh YJ, Nielsen KF, Klaenhammer TR, Barrangou R, Abou HM. *Lactobacillus acidophilus* metabolizes dietary plant glucosides and externalizes their bioactive phytochemicals. MBio. 2017;**8**:e01421-e01417. DOI: 10.1128/mBio.01421-17

[132] Basholli-Salihu M, Schuster R, Mulla D, Praznik W, Viernstein H, Mueller M. Bioconversion of piceid to resveratrol by selected probiotic cell extracts. Bioprocess and Biosystems Engineering. 2016;**39**:1879-1885. DOI: 10.1007/s00449-016-1662-1

[133] Shimoda K, Hamada H. Synthesis of β-maltooligosaccharides of glycitein and daidzein and their anti-oxidant and anti-allergic activities. Molecules. 2010;**15**:5153-5161. DOI: 10.3390/molecules15085153

[134] Heng Y, Kim MJ, Yang HJ, Kang S, Park S. *Lactobacillus intestinalis*

efficiently produces equol from daidzein and chungkookjang, short-term fermented soybeans. Archives of Microbiology. 2019;**201**:1009-1017. DOI: 10.1007/s00203-019-01665-5

[135] Koa JA, Parka JY, Kwona HJ, Ryua YB, Jeonga HJ, Parka SJ, et al. Purification and functional characterization of the first stilbene glucoside-specific β-glucosidase isolated from *Lactobacillus kimchi*. Enzyme and Microbial Technology. 2014;**67**:59-66. DOI: 10.1016/j.enzmictec.2014.09.001

[136] Decroos K, Vanhemmens S, Cattoir S, Boon N, Verstraete W. Isolation and characterisation of an equol-producing mixed microbial culture from a human faecal sample and its activity under gastrointestinal conditions. Archives of Microbiology. 2005;**183**:45-55. DOI: 10.1007/s00203-004-0747-4

[137] Tsuda H, Shibata E. Bioconversion of daidzin to daidzein by lactic acid bacteria in fermented soymilk. Food Science and Technology Research. 2017;**23**:157-162. DOI: 10.3136/fstr.23.157

[138] Lim YJ, Lim B, Kim HY, Kwon SJ, Eom SH. Deglycosylation patterns of isoflavones in soybean extracts inoculated with two enzymatically different strains of lactobacillus species. Enzyme and Microbial Technology. 2020;**132**:109394. DOI: 10.1016/j.enzmictec.2019.109394

[139] Wang XL, Kim HJ, Kang SI, Kim SI, Hur HG. Production of phytoestrogen S-equol from daidzein in mixed culture of two anaerobic bacteria. Archives of Microbiology. 2007;**187**:155-160. DOI: 10.1007/s00203-006-0183-8

[140] Burns J, Yokota T, Ashihara H, Lean MEJ, Crozier A. Plant foods and herbal sources of resveratrol. Journal of Agricultural and Food Chemistry. 2002;**50**:3337-3340. DOI: 10.1021/jf0112973

[141] Meng X, Zhou J, Zhao CN, Gan RY, Li HB. Health benefits and molecular mechanisms of resveratrol: A narrative review. Food. 2020;**9**:340. DOI: 10.3390/foods9030340

Chapter 4

Bacteriocins: Applications in Food Preservation and Therapeutics

Parul Thapar and Mohinder Kumar Salooja

Abstract

The awareness in preventing the use of chemical preservatives for food has increased. Not only this, but the prevalence of antimicrobial resistance in the food-borne pathogens that can cause infections such as food poisoning is also at a rise. This has led in the growing demand for the safe food. The bacteriocins can be used as an effective alternative in food preservation and safety. Bacteriocins are ribosomally synthesized proteins that possess certain inhibitory activities against diverse group of undesirable microorganisms. These are produced by both Gram-positive and Gram-negative bacteria and some of the archaeal species. Bacteriocins are safe for human consumption, since they can be degraded by proteolytic enzymes in the gastrointestinal tract. In this chapter, focus is made on an alternative and safe approach for food preservation and therapeutics through bacteriocins. The applications of different types of bacteriocins in preserving food are mentioned with regard to increased shelf life, additives, and packaging. Not only this, but also bacteriocins benefit in boosting the immune system and possess certain anticancer properties. Bacteriocins can also be used in controlling the antimicrobial resistance in certain food-borne pathogens. They are the future antimicrobial proteins for the food preservation and therapeutics in a cost-effective manner.

Keywords: bacteriocins, shelf life, food preservation, immune system, *Lactobacillus* spp., therapeutics, antimicrobial resistance

1. Introduction

According to the present scenario, where people have become health and diet conscious, the use of chemical preservatives in foods has become a concern in a healthy lifestyle [1]. Another concern is the antimicrobial resistance of the food-borne pathogens within the food that can lead to spread of infections such as food poisoning [2].

Antimicrobial resistance is the ability of the microorganisms (bacteria, protozoa, fungi, or virus) to continue to grow even when they are exposed to antimicrobial medicines that are meant to kill or inhibit their pathogenic activities. As a result, medicines become ineffective and a person is not cured. It mainly happens due to overuse of the same medicine against a specific disease, which let the genes of the microbes to get adapted to a particular medicine [2]. Antimicrobial resistance is increased in various microorganisms that can lead to food spoilage and cause severe infections. The antimicrobial resistance of these food-borne pathogens in food-producing animals can be spread to humans *via* contaminated food or water and also through direct contact with the animals [3]. This focuses on "One- Health Concept," which is an approach that recognizes "the health of people is closely connected to the health of an animal" [4].

In one of the studies conducted by European Union, it is shown that the *Salmonella* spp. isolated from turkeys, meat, and pork showed antimicrobial resistance to the drugs such as sulfonamides, tetracycline, ampicillin, and fluoroquinolones. Also, the species of *Escherichia coli* isolated from meat and turkey showed antimicrobial resistance to sulfonamides, tetracycline, and ampicillin drugs against the patients suffering from food poisoning after consumption of these animal foods [5].

According to the Centre for Disease Control and Prevention (CDC), the antimicrobial resistance of pathogens from the family Enterobacteriaceae, including *Escherichia coli*, *Shigella,* and *Salmonella* spp., posts a serious threat to the world [4].

The increase in the demand of natural preservatives to be used in food products and natural sources that can inhibit antimicrobial resistance has led the researchers to think about different approaches toward food preservation and safety. Therefore, the application of bacteriocins can be an effective alternative.

Bacteriocins are ribosomally synthesized antimicrobial proteins [6]. These are produced by both Gram-positive and Gram-negative bacteria and some of the archaeal species. They possess certain inhibitory activities against diverse group of undesirable microorganisms [7].

The bacteriocins from Gram-positive bacteria show the following characteristics—antimicrobial in action, narrow spectrum, active against relative species of organisms, and in broad spectrum, active against both Gram-positive and Gram-negative organisms and some fungi [8]. The large group of microbial species producing bacteriocins is mainly the lactic acid bacteria (LAB). These are a group of Gram-positive, non-spore forming, non-motile, non-respiring bacteria, which produce a variety of antimicrobial compounds such as lactic acid, acetic acid, ethanol, formic acid, fatty acids, hydrogen peroxide, and bacteriocins [9]. The genera includes *Lactobacillus*, *Leuconostoc*, *Pediococcus*, *Streptococcus*, *Alloiococcus*, *Carnobacterium*, *Dolosigranulum*, *Enterococcus*, *Oenococcus*, *Tetragenococcus*, *Vagococcus*, *Weissella* [10].

Bacteriocins can be considered as an alternative in food preservation compared to chemical preservatives. These are safe for human consumption, since they can be degraded by proteolytic enzymes in the gastrointestinal tract [6]. Bacteriocins become inactive when they come in contact with the digestive enzymes in the stomach, as the enzymes such as pepsin denature the bacteriocins [9]. Also, presently, 50 LAB strains have obtained the Qualified Presumption of Safety (QPS) status by the European Food Safety Agency [11].

2. Classification of bacteriocins

Different types of bacteriocins have been classified according to size, inhibitory mechanism, target cells, spectrum of action, interaction with immune system, and biochemical features [12]. The bacteriocins have different mechanisms of action: bactericidal, with or without cell lysis through cell wall, and bacteriostatic, inhibiting the cell growth by inhibiting gene expression or protein production [6]. Accordingly, the types of bacteriocins are represented in **Table 1**.

Class	Typical producing species	Properties	Examples	Mode of action	References
I	*Lactococcus lactis* sub sp. *Lactis*	Contain unique amino acids, that is, lanthionine and methyllanthionine;<5 kDa	Nisin, Lactocin, Mersacidin	Bactericidal; by targeting the peptidoglycan layer of the bacterial cell wall. Except nisin—targets by both mechanisms	[12–14]
			Some novel lanthionines—linardin, azoline, cyanobactin, glycocin, , thiopeptide, and lasso peptide	Bactericidal	[15]
IIa	*Leuconostoc gelidum*	Heat stable, non-modified, cationic, hydrophobic peptides; contain a double–glycine leader peptide; pediocin-like peptides; <10 kDa	Pediocin PA1, Sakacin A, Leucocin A		[13, 16, 17]
IIb	*Enterococcus faecium*	Require synergy of two complementary peptides; mostly cationic peptides	Lactococcin G, Plantaricin A, Enterocin X		[18, 19]
IIc	*Lactobacillus acidophilus*	Affects membrane permeability and cell wall formation	Acidocin B, Enterocin P, Reuterin 6		[20, 21]
III	*Lactobacillus helveticus* from Swiss cheese	Heat-labile; large molecular mass peptides; >30 kDa	Lysostaphin, Enterolysin A, Helveticin J	Bacteriostatic	[13–15, 19, 21]

Table 1.
Classification of bacteriocins.

3. Applications of bacteriocins in food preservation

Bacteriocins have been used in biopreservation of various foods, either alone or in combination with other methods of preservation, such as hurdle technology [19]. The criteria for the bacteriocins to be used for food preservation or food safety are [22, 23]:

- Bacteriocin-producing strains should be food grade (GRAS or QPS).
- Exhibit a broad spectrum of inhibition.

- Show high specific activity.
- Have no health risks.
- Beneficial effects on foods (e.g., improve safety) and do not affect quality and flavor of food.
- They should be heat and pH stable.

Bacteriocins can be incorporated in foods either by inoculating directly into the food or incorporating in food packaging films/coatings, which will improve their activity or stability in complex food systems [24].

To protect the foods from contamination with certain microorganisms such as *Listeria monocytogenes* (pathogens in cheese) and *Streptococcus aureus* in dairy products [25] while in meat and fermented sausages, contamination of *Clostridium botulinum* [26–28] and other bacterial pathogens such as Campylobacter spp.,

S.No.	Bacteriocin-producing strains	Products incorporated	Active against	References
1.	*Lactococcus lactis* 8L1A and 8L1 B	Starters for cheese	Bacterial pathogens	[29–31]
2.	*Lactobacillus sakei* sub sp. *sakei* 2a	Cheese spread	*Listeria monocytogenes*	[21, 32]
3.	*Lactiplantibacillus plantarum* CCDM1078	Cheese spread	*Listeria monocytogenes*	
4.	*Staphylococcus equorum* SE3	Cheese	*Listeria monocytogenes*	[33]
5.	*Enterococcus faecium*	Fresh cheese whey	*Listeria monocytogenes*	[34]
6.	*Lactococcus lactis* (Nisin producer)	Fresh cheese	*Listeria monocytogenes*	[35]
7.	Enterocin from *Enterococcus faecalis* LBB1K3	Model fresh cheese	*Listeria monocytogenes*	[36]
8.	Lacticin 481 (*Lactococcus lactis*)	Fresh cheese	*Listeria monocytogenes*	
9.	Semi -purified pediocin	Fermented cheese	*Staphylococcus aureus*	[37]
10.	Semi-purified lacticin	Fresh cheese	*Listeria monocytogenes*	
11.	Semi-purified bacteriocin—BacFL31	Turkey meat	*Listeria monocytogenes* and *Salmonella typhi*	
12.	*Enterococcus faecium* KEB2	Raw and sterile milk	*Listeria monocytogenes*	
13.	Aureocin A70 (*Staphylococcus aureus*)	UHT skim milk	*Listeria monocytogenes*	[38]
14.	Lactococcin BZ (*Lactococcus lactis*)	Skim and UHT milk	*Listeria monocytogenes*	[39]
15.	*Leuconostoc mesenteroides*, Leucocin K7	UHT and whole-fat milk	*Listeria monocytogenes*	[40–42]

Table 2.
Bacteriocins-producing strains used as biopreservatives.

S. No.	Bacteriocin-producing strains	Food products	Active against	References
1.	Nisin	Cured meat products	*Listeria monocytogenes*, *Clostridium spp.*, *Bacillus cereus*, *Escherichia coli*, and Salmonella spp.	[43]
2.	Plantaricin 423	Cheese	*Listeria innocua*	[44]
3.	Nisin A, nisin Z and lacticin-481	Cottage cheese	*Listeria monocytogenes*	[45, 46]
4.	Nisin A (Nisapsin^R)	Milk pudding	Spore formers	[47]
5.	Nisin A (Nisapsin^R)	Cheese	*Staphylococcus aureus*	[48]
6.	*Lactococcus lactis* N564, Nisin	Cow milk	*Listeria monocytogenes and Staphylococcus aureus*	[49]

Table 3.
Some commercialized bacteriocins.

Salmonella spp., *Escherichia coli*, *Listeria monocytogenes*, *Staphylococcus aureus*, *Yersinia enterocolitica* [29]; some viral and parasitic species such as *Toxoplasma gondii* and *Taenia solium* [30], certain immobilized form of bacteriocins have been developed as antimicrobial packaging films [26–28].

3.1 Bacteriocins as biopreservatives

Some of the bacteriocin-producing strains are used as biopreservatives in food preservation but not marketed yet. These are shown in **Table 2**. There are other bacteriocins that have been proposed for industrial applications such as Enterocin AS-48 [26] and Lacticin 347 [41].

Till present, there are some bacteriocins that have been commercialized as food additives including nisin, with trade name Nisaplin andDanisco; pediocin A1, with trade name Microgard^{TM} and ALTA2431 [42]. The other commercialized bacteriocins active against certain food spoilage microorganisms are represented in **Table 3**.

3.2 Bacteriocins as food additives

Food additives are the substances added to food to maintain or improve the safety, freshness, taste, texture, or appearance. Food additives need to be checked for potential harmful effects on human health before they can be marketed [50]. Some of the bacteriocins that are applied as food additives are presented in **Table 4**.

S.No.	Bacteriocin-producing strains	Food additives	References
1	Bacteriocin-producer adjunct of *Pediococcus acidilacti*	Cheddar and semi-hard cheeses	[44]
2.	Bacteriocin-producing adjunct of *Lactobacillus acidophilus*	Food fermentation	[49]

Table 4.
Bacteriocins as food additives.

Bacteriocin-producing strains	Food packaging	Features	References
Nisin coated on polyethylene films	Packaging of poultry	Reduced risk of *Salmonella* spp.	[51]
Nisin absorbed onto salinized silica surfaces	Packaging of meat	Inhibited *Listeria monocytogenes*	
Lacticaseibacillus casei producing bacteriocin strain	The bacteriocins adsorbed on the packages on cheese, diffused through the medium, thus, inhibiting the growth of undesirable organisms like *Escherichia coli* and *Staphylococcus aureus*	Reduced the risk of pathogen development and extending the shelf-life of foods.	[21, 52]

Table 5.
Bacteriocins in food packaging.

3.3 Bacteriocins in food packaging

Antimicrobial packaging film prevents microbial growth on food surface by direct contact of the package with food surfaces such as meat and cheese [51]. In one of the studies, bacteriocin-producing lactic acid bacteria were isolated from Yakult (a probiotic drink) to develop an antimicrobial packaging (**Table 5**) [52].

4. Applications of bacteriocins as therapeutics

4.1 Bacteriocins in boosting immune system

Bacteriocins play an important role in boosting the immune system of the human beings. They allow the survival of specific bacterial strains in the

S. No.	Bacteriocins	Immune system, disease prevention, and Anticancer properties	Reference
1.	Bactofencin A or bacteriocin 21 produced by *Enterococcus faecalis*	Kill multidrug-resistant bacteria	[53]
2.	Pyocin	Control *Pseudomonas* lung infections in patients with cystic fibrosis.	
3.	Colicin A and Colicin E1	Inhibitory activity against the growth of eleven different tumor cell lines	
4.	Colicin D and Colicin E2	Inhibitory effect against murine leukemia cells P388	
5.	Colicin E3	Suppressed the malignant transformation of a chicken monoblast line	
6.	Bacteriocins of *Escherichia coli*	Act against human colorectal carcinoma cells	
7.	Nisin (commercial)	Act against human colorectal carcinoma cells	
8.	Bacteriocins from lactic acid bacteria	Inhibits vancomycin-resistant *Enterococci*, *Salmonella enteritidis*, *Clostridium difficile*, and *Listeria monocytogenes*	[54]

Table 6.
Bacteriocins in immune system.

S.No.	Bacteriocins or producer strains	AMR-resistant strains	Reference
1.	Nisin	Methicillin-resistant *Staphylococcus aureus* (MRSA) in goat milk	[55]
2.	Lip A	*Pseudomonas aeruginosa*	[56, 57]
3	BAC-1B-17 from *Bacillus subtilis* (thermostable) and Sonorensin from *Bacillus sonorensis*	Methicillin-resistant *Staphylococcus aureus* (MRSA)	[58, 59]
4.	Strains of *Lactobacillus helveticus*, *Lactobacillus delbreuckii*, *Lactococcus lactis*	Vancomysin-resistant *Salmonella enetritidis*, *Clostridium difficle*, *Listeria monocytogenes*	[60, 61]

Table 7.
Bacteriocins in inhibiting antimicrobial-resistant (AMR) food pathogens.

gastrointestinal tract. For example, some strains of *Lactobacillus* spp. are able to resist modification by the host diet; also inhibit other killing factors and colonization by pathogenic species within the intestine. Hence, they improve the gut barrier function and host immune response. Bacteriocins also possess anticancer properties by inhibiting the growth of the tumor cells in some animals and humans. The examples of bacteriocins in immune system and anticancer properties are mentioned in **Table 6**.

4.2 Bacteriocins in inhibiting antimicrobial-resistant (AMR) food pathogens

The bacteriocins inhibiting the antimicrobial-resistant strains of the microbes present within the food work on the principle of quorum sensing. They inhibit the competitive strains by directly influencing the niche competition [55]. The bacteriocins or producing strains that can inhibit the antimicrobial-resistant food pathogens are mentioned in **Table 7**.

5. Conclusion and future scope

Bacteriocins are the antimicrobial proteins, which could be categorized as antibiotics, but they are not. The major difference between bacteriocins and antibiotics is that the bacteriocins are species-specific and their activity is restricted to a particular strain of species; on the other hand, antibiotics have a wider activity spectrum. The bacteriocins are produced by the lactic acid bacterial species. The increased antimicrobial resistance and growing awareness of microbiome for the importance of human health underscore the need of this class of antimicrobials, as an approach for the treatment of infectious diseases spread by the antimicrobial-resistant food-borne pathogens such as *Salmonella* spp., *Listeria* spp. etc. [62]. Thus, these bacteriocins can be a part of sustainable development goals by delivering safe foods with longer shelf life [1]. The bacteriocins are the future antimicrobial proteins for the preservation of food and as therapeutics even in a cost-effective manner. The species of Enterococci have been identified as antimicrobials against vancomycin-resistant pathogens of the food industry [63, 64]. The food and dairy industries and healthcare sector should be more focused on the use of bacteriocins for food preservation and as therapeutics in cancer treatment, respectively.

Author details

Parul Thapar[1*] and Mohinder Kumar Salooja[2]

1 Department of Food Science and Technology, School of Science, Gandhi Institute of Technology and Management (GITAM) Deemed University, Hyderabad, India

2 School of Agriculture, Indira Gandhi National Open University, New Delhi, India

*Address all correspondence to: parul.thapar@yahoo.com

References

[1] Rossi F, Pallotta LM. Bacteriocin producing cultures: A sustainable way for food safety improvement and probiotics with additional health promoting effects. International Journal of Medical Biology Frontiers. 2019;**22**:59-91

[2] World Health Organization. Antibiotic Resistance: Key Facts. 2020. Available from: https://www.who.int/en/news-room/fact-sheets/detail/antibiotic-resistance

[3] Oniciuc EA, Likotrafiti E, Alvarez MA, Prieto M, Lopez M, Alvarez OA. Food processing as a risk factor for antimicrobial resistance spread along the food chain. Current Opinion in Food Science. 2019;**30**:21-26

[4] Centres for Disease Control and Prevention. Available from: https://www.cdc.gov/onehealth/basics/index.html

[5] Dall C. EU Notes Rising Resistance in Common Foodborne Pathogens. News Report. 2020. Available from: https://www.cidrap.umn.edu/news-perspective/2020/03/eu-notes-rising-resistance-common-foodborne-pathogens

[6] Silva CCG, Silva SPM, Ribeiro SC. Application of Bacteriocins and Protective Cultures in Dairy Food Preservation. Frontiers in Microbiology. 2018;**9**:1-15

[7] Lopetuso LR, Giorgio ME, Saviano A, Scaldaferri F, Gasbarrini A, Cammarota G. Antimicrobial activity of bacteriocins of Lactic Acid Bacteria on Listeria monocytogenes, Staphylococcus aureus and Clostridium tyrobutyricum in cheese production. Journal of International Molecular Science. 2019;**20**(183):1-12

[8] Salazar MEL, Galan WLJ, Moreno MVR, Reyes MA, Pereyra AB. Bacteriocins synthesized by Bacillus thuringiensis: Generalities and potential applications. Review in Medical Microbiology. 2016;**27**(3):95-101

[9] Upendra SR, Khandelwal P, Jana K, Kumar AN, Devi G. Bacteriocin Production from Indigenous Strains of Lactic Acid Bacteria Isolated from Selected Fermented Food Sources. International Journal of Pharma Research and Health Sciences. 2016;**4**(1):982-990

[10] Quinto EJ, Jimenez P, Caro I, Tejero J, Mateo J, Girbes T. Probiotic Lactic Acid Bacteria: A Review. Food and Nutrition Science. 2014;**5**:1765-1775

[11] Ricci A, Allende A, Bolton D, Chemaly M, Davies R, Girones R. Listeria monocytogenes contamination of ready-to-eat foods and the risk for human health in the EU. Journal of EFSA. 2017;**1**:15-33

[12] Hegarty JW, Guinane CM, Ross RP, Hill C, Cotter PD. Recent advances in microbial fermentation for dairy and health. Journal of EFSA. 2016;**5**:2587

[13] Parada JL, Caron CR, Medeiros ABP, Soccol CR. Bacteriocins from lactic acid bacteria: purification, properties and use as biopreservatives. Brazilian Archives of Biology and Technology. 2007;**50**:521-542

[14] Todorov SD. Bacteriocins from Lactobacillus plantarum – production, genetic organization and mode of action. Journal of Brazilian Microbiology. 2009;**40**:209-221

[15] Mahrous H, Mohamed A, El-Mongy MA, El-Batal AI, Hamza HA. Effect of Probiotic Chocolate in the Reduction of Streptococcus Mutans Count. Food and Nutrition Sciences. 2013;**4**:342-356

[16] Umu OC, Bauerl C, Oostindjer M, Pope PB, Hernandez PE, Perez-Martinez G, et al. Lactic acid bacteria. PLoS One. 2016;**1**:11-13

[17] Zacharof MP, Lovitt RW. Bacteriocins produced by Lactic acid bacteria. APCBEE Procedia. 2012;**2**:50-56

[18] Hu CB, Malaphan W, Zendo T, Nakayama J, Sonomoto K. Enterocin X, a novel two-peptide bacteriocin from enterococcus faecium. Applied and Environmental Microbiology. 2010;**76**:4542-4545

[19] Perez RH, Zendo T, Sonomoto K. Novel bacteriocins from lactic acid bacteria (LAB): Various structures and applications. Microbial Cell Factories. 2014;**13**:3-6

[20] Suskovic J, Kos B, Beganovic J, Pavunc AL, Habjanic K, Matosic S. Antimicrobial Activity – The Most Important Property of Probiotic and Starter Lactic Acid Bacteria. Food Technology and Biotechnology. 2010;**48**:296-307

[21] Zheng J, Salvetti E, Franz MAPC, Mattarelli P. A taxonomic note on the genus Lactobacillus: Description of 23 novel genera, emended description of the genus Lactobacillus Beijerinck 1901, and union of Lactobacillaceae and Leuconostocaceae. International Journal of System Evolution and Microbiology. 2020;**70**:2782-2858

[22] Cotter PD, Hill C, Ross RP. Nature Reviews. Microbiology. 2005;**3**:777-788

[23] Leroy F, De Vuyst L. Bacteriocins of lactic acid bacteria to combat undesirable bacteria in dairy products. Journal of Australian Dairy Technology. 2010;**65**:143-149

[24] Salgado PR, Ortiz CM, Musso YS, Di GL, Mauri AN. Edible films and coatings containing bioactives. Current Opinion in Food Science. 2015;**5**:86-92

[25] Melo J, Andrew P, Faleiro M. Listeria monocytogenes in cheese and the dairy environment remains a food safety challenge: The role of stress responses. Bacteriocins: developing innate immunity for food. Journal of International Food and Research. 2015;**67**:75-90

[26] Sanchez HM, Montalban LM, Cebrian R, Valdivia E, Martinez BM, Maqueda M. AS-48 bacteriocin: Close to perfection. Cellular and Molecular Life Sciences. 2011;**68**:2845-2857

[27] Ibarguren C, Celiz G, Diaz AS, Bertuzzi MA, Daz M, Audisio MC. Gelatine based films added with bacteriocins and a flavonoid ester active against food-borne pathogens. Innovative Food Science and Emerging Technologies. 2015;**28**:66-72

[28] Narsaiah K, Wilson RA, Gokul K, Mandge H, Jha S, Bhadwal S. Effect of bacteriocin-incorporated alginate coating on shelf-life of minimally processed papaya (Carica papaya L.). Postharvest Biology and Technology. 2015;**100**:212-218

[29] Tack D, Ray L, Griffin P. Preliminary Incidence and Trends of Infections with Pathogens Transmitted Commonly Through Food – Foodborne Diseases Active Surveillance Network, 10 U.S. Sites, 2016-2019. Morbility Mortalality Weekly Report. 2020;**69**:509-514

[30] Franssen F, Gerard C, Cozma-PetruÛ A, Vieira-Pinto M, Jambrak AR, Rowan N, et al. Inactivation of parasite transmission stages: Efficacy of treatments on food of animal origin. Trends in Food Science and Technology. 2019;**83**:114-128

[31] Cotelo MF, Schein KP, Salvo SSG, Abirad PMZ, Techera SBC. Antimicrobial properties of lactic acid bacteria isolated

from uruguayan artisan cheese. Food Science and Technology. 2013;**33**(4):1-3

[32] Patrovsky M, Kourimska L, Havlikova S, Markova J, Pechar R, Rada V. Utilization of bacteriocin-producing bacteria in dairy products. Mljekarstvo. 2016;**66**:215-224

[33] Bockelmann W, Koslowsky M, Goerges S, Scherer S, Franz C, Heller KJ. Growth inhibition of Listeria monocytogenes by bacteriocin-producing Staphylococcus equorum SE3 in cheese models. Food Control. 2017;**71**:50-56

[34] Aspri M, Oconnor PM, Field D, Cotter PD, Ross P, Hill C. Application of bacteriocin-producing Enterococcus faecium isolated from donkey milk, in the bio-control of Listeria monocytogenes in fresh whey cheese. Journal of International Dairy Research. 2017;**73**:1-9

[35] Khan I, Oh DH. Integration of nisin into nanoparticles for application in foods. Innovative Food Science & Emerging Technologies. 2016;**34**:376-384

[36] Ribeiro SC, Ross RP, Stanton C, Silva CC. Characterization and application of antilisterial enterocins on model fresh cheese. Journal of Food Proteins. 2017;**80**:1303-1316

[37] Vandera E, Lianou A, Kakouri A, Feng J, Koukkou AI, Samelis J. Enhanced control of Listeria monocytogenes by Enterococcus faecium KE82, a multiple enterocin–producing strain, in different milk environments. Journal of Food Proteins. 2017;**80**:74-85

[38] Fagundes PC, De FFM, Da Silva SOC, Da Paz JAS, Ceotto VH, Alviano DS. The four-component aureocin A70 as a promising agent for food biopreservation. Journal of International Food Microbiology. 2016;**237**:39-46

[39] Yildirim Z, Oncul N, Yildirim M, Karabiyikli S. Application of lactococcin BZ and enterocin KP against Listeria monocytogenes in milk as biopreservation agents. Acta Alimentaria. 2016;**45**:486-492

[40] Shi F, Wang YW, Li YF, Wang XY. Mode of action of leucocin K7 produced by Leuconostoc mesenteroides K7 against Listeria monocytogenes and its potential in milk preservation. Biotechnology Letters. 2016;**38**:1551-1557

[41] Suda S, Cotter PD, Hill C, Paul RR. Lacticin 3147-biosynthesis, molecular analysis, immunity, bioengineering and applications. Current Protein & Peptide Science. 2012;**13**:193-204

[42] Simha BV, Sood S, Kumariya R, Garsa AK. Simple and rapid purification of pediocin PA-1 from Pediococcus pentosaceous NCDC 273 suitable for industrial application. Microbiological Research. 2012;**167**:544-549

[43] Fraqueza MJ, Patarata L. Laukova A. Fermented meat products health aspects. Fermented Meat Products: Health Aspects. 2017;**572**

[44] Mills S, Serrano LM, Griffin C. Inhibitory activity of Lactobacillus plantarum LMG P-26358 against Listeria innocua when used as an adjunct starter in the manufacture of cheese. Microbial Cell Factories. 2011;**10**:1-5

[45] Dal BB, Cocolin L, Zeppa G, Field D, Cotter PD, Hill C. Technological characterization of bacteriocin producing Lactococcus lactis strains employed to control Listeria monocytogenes in cottage cheese. Journal of International Food Microbiology. 2012;**153**:58-65

[46] Arques JL, Rodríguez E, Langa S, Landete JM, Medina M. Antimicrobial activity of lactic acid bacteria in dairy

products and gut: effect on pathogens. Journal of International Biomedical Research. 2015;**1**:1-5

[47] Oshima S, Hirano A, Kamikado H, Nishimura J, Kawai Y, Saito T. Nisin A extends the shelf life of high-fat chilled dairy dessert, a milk based pudding. Journal of Applied Microbiology. 2014;**116**:1218-1228

[48] Felicio BA, Pinto MS, Oliveira FS, Lempk MW, Pires ACS, Lelis CA. Effects of nisin on Staphylococcus aureus count and physicochemical properties of Minas Frescal cheese. Journal of Dairy Science. 2015;**98**:4364-4369

[49] Alves FCB, Barbosa LN, Andrade B, Albano M, Furtado FB, Pereira AFM. Short communication: inhibitory activities of the lantibiotic nisin combined with phenolic compounds against Staphylococcus aureus and Listeria monocytogenes in cow milk. Journal of Dairy Science. 2016;**99**:1831-1836

[50] World Health Organization. Food Additives. 2018. Fact Sheet. Available from: https://www.who.int/news-room/fact-sheets/detail/food-additives

[51] Deshmukh PV, Thorat PR. Bacteriocins: a new trend in antimicrobial food packaging. International Journal of Advanced Research in Engineering and Applied Science. 2013;**2**(1):1-12

[52] Damania P, Patel R, Shaw R, Kataria RP, Wadia A. Development of antimicrobial packaging materials for food preservation using bacteriocin from Lactobacillus casei. Microbiological Research. 2016;**7**(1):19-22

[53] Yang SC, Lin CH, Sung CT, Fang JY. Antibacterial activities of bacteriocins: Application in foods and pharmaceuticals. Frontiers in Microbiology. 2014;**5**:241

[54] Joo NE, Ritchie K, Kamarajan P, Miao D, Kapila YL. Nisin, an apoptogenic bacteriocin and food preservative, attenuates HNSCC tumorigenesis via CHAC1. Cancer Medicine. 2012;**1**:295-305

[55] Kommineni S, Bretl DJ, Lam V, Chakraborty R, Hayward M, Simpson P, et al. Bacteriocin production augments niche competition by enterococci in the mammalian gastrointestinal tract. International Journal of Food Microbiology. 2015;**526**:719-722

[56] Perales AJ, Rubino S, Martinez M. LAB bacteriocins controlling the food isolated (drug-resistant) staphylococci. Frontiers in Microbiology. 2018;**9**:1143

[57] Escolanoa MR, Cebrianb R, Escolanoa JM, Rosalesa MJ, Maquedab M, Morenoa SM, et al. Insights into Chagas treatment based on the potential of bacteriocin AS-48. Drugs Drug Resist. 2019;**10**:1-8

[58] Ansari A, Zohra RR, Tarar OM, Qader SAU, Aman A. Characterization, Cytotoxic Analysis and Action Mechanism of Antilisterial Bacteriocin Produced by Lactobacillus plantarum Isolated from Cheddar Cheese. BMC Microbiology. 2018;**18**(1):1-10

[59] Chopra L, Singh G, Jena KK, Sahoo DK. Sonorensin: A new bacteriocin with potential of an anti-biofilm agent and a food biopreservative. Scientific Reports. 2015;**5**:13412

[60] Umu OCO, Rudi K. Diep DB. Modulation of the gut microbiota by prebiotic fibres and bacteriocins. Microbial Ecology in Health and Disease. 2019;**28**

[61] Rea MC, Alemayehu D, Ross RP, Hill C. Gut solutions to a gut problem: bacteriocins, probiotics and bacteriophage for control of Clostridium

difficile infection. Journal of Medical Microbiology. 2013;**62**(9):1369-1378

[62] Laranjo M, Potes EM, Elias M. Role of Starter Cultures on the Safety of Fermented Meat Products. Frontiers in Microbiology. 2019;**10**(853):1-11

[63] Grujovic M, Mladenovic K, Petrovic TZ. Comic, L. Assessment of the antagonistic potential and ability of biofilm formation of Enterococcus spp. isolated from Serbian cheese. Veterinarski arhiv. 2019;**89**(5):653-667

[64] Suarez N, Weckx S, Minahk C, Hebert EM. Saavedra L. Metagenomics-based approach for studying and selecting bioprotective strains from the bacterial community of artisanal cheeses. International Journal of Food Microbiology. 2020;**335**:108894

Chapter 5

Use of *Lactobacillus* for Lactic Acid Production from Agro-Industrial By-Products

Ederson Freire-Almeida and Pedro Maldonado-Alvarado

Abstract

Agro-industrial by-products have not been efficiently valorized. *Lactobacillus* used to transform these by-products into interesting metabolites is a way to increase the adding-value of these residues and to contribute to the circular economy. These lactic acid bacteria (LAB) metabolize the available substrate produced by enzymes that are responsible for breaking complex carbohydrates into glucose and subsequently obtaining lactic acid through glycolysis in a homofermentative process. By-products used like substrates to produce lactic acid must be rich in carbohydrates e.g. whey, cassava peel, pineapple peel, and molasses, among others. In addition, from lactic acid obtained, it is possible to develop functional foods such as easily-assimilated beverages and to be antagonists to pathogenic microorganisms such as *E. coli*, improve the quality of final products and extract compounds of interest like pigments.

Keywords: *lactobacillus*, lactic acid, agro-industrial by-products, circular economy, lactic acid bacteria (LAB)

1. Introduction

Agribusiness is a productive activity that combines the industrial and agricultural processes to obtain products with added value, which can have food or non-food applications [1]. As a result of the agro-industrial activity, waste is generated in significant quantities that represent an environmental problem, due to the inefficiency of its use or the lack of knowledge about appropriate methods for its treatment [2]. Agro-industrial waste is defined as a solid or liquid material product of the processing of primary products, which are not used in the production process. However, these can be used to generate another product of greater economic or ecological value [2].

The use of agro-industrial by-products obeys the guidelines of the circular economy, where the concept of waste disappears, and this element will be used as a resource for nature, society, and industry. The circular economy aims to avoid the creation of waste that can have a negative impact on the environment, climate, and health [3]. In addition, the use of agro-industrial by-products may contribute to the Sustainable Development Goals (SDGs) of The United Nations Development Program (UNDP), in particular, "no poverty" and "zero hanger".

The use of LAB to valorize food by-products through the production of lactic acid, has generated interest in recent years. LAB are microorganisms that are widely used in the food industry since they allow obtaining fermented products with pleasant sensory characteristics. In addition, to favor the intestinal biota, they have been attributed properties such as the ability to remove toxic metals from aqueous solutions [4], generate unfavorable environments for the growth and development of pathogenic microorganisms, such as *Escherichia coli*, *Listeria monocytogenes*, *Pseudomonas*, etc. [5]. Also, LAB as *Lacticaseibacillus* and *Lactobacillus acidophilus* strains have a positive in vivo response to reduce the bioavailability of methyl mercury (CH_3Hg) in the human being [6].

LAB have been classified as Gram-positive bacteria. These microorganisms are facultative anaerobes, they do not form spores and have the shape of bacilli. The main metabolite generated for LAB in the fermentation process is lactic acid, an organic acid very interesting in industry [7]. LAB are demanding microorganisms in terms of nutritional requirements, they need certain amino acids and vitamins of the B complex for their development. The most common media used in the fermentation of LAB include, sugar (glucose, lactose, or sucrose), calcium carbonate (used as a buffer for the medium), and malt germ (for the contribution of nitrogenous elements and growth factors). The optimal fermentation temperature for *Lactobacillus* is between 15 and 55° C [8] and the fermentation time should be around 6 days, to achieve around a yield of 90% in lactic acid [9].

Lactic acid production may result from homolactic or heterolactic metabolic routes. The homolactic route consists of the conversion of glucose to pyruvic acid, which through the presence of lactate dehydrogenase, acts as an electron acceptor in the oxidation of NADH, becomes lactic acid. Heterolactic fermentation is carried out through the pentose pathway where there is the formation of xylulose-5-phosphate, an intermediate product in the formation of lactic acid [10]. Finally, other products, in addition to lactic acid, are formed such as ethanol, acetic acid, and carbon dioxide [9].

Lactic acid is composed of two functional groups: carboxyl and alcohol, obtaining an asymmetric carbon that provides it the optical activity. This acid has two optical isomers: lactic L (+) and lactic D (−) (**Figure 1**). However, only the L (+) isomer, which is considered a GRAS substance by the FDA, is considered a food additive e.g., as acidulant and preservative. However, it has other industrial importance in cosmetic, pharmaceutical, and chemical applications [12].

Lactic acid has been widely used in recent decades, for example as the precursor of polylactic acid (a biodegradable biopolymer to manufacture packaging material for the food industry), 3D printing applications, and for medical uses, among others. Currently, the production of lactic acid is achieved through fermentation, however,

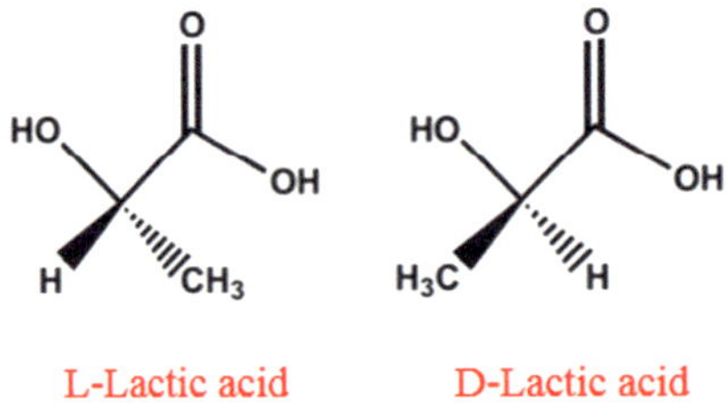

Figure 1.
Isomeric configurations of lactic acid [11].

the use of raw materials as sources of carbon and energy represents an environmental problem with a high cost. For this reason, environmentally friendly alternatives have been evaluated from agro-industrial by-products, as raw materials to obtain this acid [13]. For example, the use of whey cassava bark and pineapple bark, using immobilized *Lactobacillus delbrueckii subsp. delbrueckii* [14], Bioconversion of agroindustrial co-products into lactic acid by *Latilactobacillus sakei* [15]. Efficient conversion of agro-industrial waste into D (−) lactic acid with *L. delbrueckii subsp. delbrueckii* [16], the use of *Lactobacillus*, in bakery by-products, meat, whey, etc. [17]. In this chapter, the use of agro-industrial by-products is exposed through the use and application of LAB such as *Lactobacillus* in order to provide adding value to these interesting substrates.

2. Circular economy in agroindustry

In recent decades, the exploitation of natural resources and raw materials has increased, due to economic, productive, and environmental problems. These factors are decisive for the circular economy, which has forced it to change from a linear model to a circular model [18]. A linear economic model is based on the production, use, and disposal of a good or product. In the context of environmental sustainability, the circular economy proposes a strategy that reduces the negative impact on the environment where the final product is the source of value creation, increasing its useful life [19]. Circular economy is based on key strategies such as waste prevention design, where both, products and services are designed and created so that the production of waste is minimal or eliminated. Likewise, the idea is to give a second life to a product and thus reduce raw material to produce a new product [3]. Thus, the products must be more versatile, modular, and simple to adapt to different applications during their shell-life. In addition, it is important to use renewable energy to reduce the negative impact on the environment [3].

The benefits of the circular economy are linked to higher economic growth by using resources efficiently, creation of employment opportunities through the creation of industries that promote innovation and entrepreneurship, net savings in the cost of raw materials, reduction of environmental pollution in the form of carbon dioxide emission, reduction of water pollution and responsible use of the soil [20]. In 2015, the United Nations proposed the fulfillment of 17 Sustainable Development goals by 2030. To meet these objectives, there must be a joint action of governments, the private sector, civil society, etc. In the case of companies, it consists of promoting business models that are more committed to ecosystems, society, and living conditions of the world population [21].

3. Valorization of agro-industrial by-products using *lactobacillus*

At present, cellulosic materials are attractive as possible replacements for edible raw material made up of starch [22], materials such as agricultural residues are considered inexpensive and captivating materials to produce lactic acid since they are a great source of carbohydrates [23, 24]. The most used LAB to produce lactic acid by fermentation are *Lactiplantibacillus plantarum, Amylolactobacillus. amylophilus, L, delbrueckii subsp. delbrueckii, L. delbrueckii subsp. bulgaricus* and *Lactobacillus acidophilus* [25–28]. For optimal lactic acid production it is necessary to make several considerations: strain of the microorganism with which to work, carbon source, temperature, pH, incubation time, and recycling of cells by immobilization [29]. For this reason, many attempts have been made to find low-priced substrates and interest has arisen in recycling agro-industrial by-products such as wheat bran, distillery residues, beer production residues, etc. These materials that are no longer used in different industries, could reduce the cost of lactic acid production, since raw materials such as starch and lignocellulosic materials require a physicochemical or enzymatic treatment for fermentations and that LAB can directly take advantage of these substrates [22].

For valorization of agro-industrial by-products using *Lactobacillus*, e.g., the production of lactic acid, several studies have been carried out with different types of substrates. For example, bakery waste with the microorganism *Thermoanaerobacterium aotearoense* [30], using whey with *S. thermophillus* and *L. delbrueckii subsp. delbrueckii* [31]. In addition, using cane bagasse as a substrate and *Lactiplantibacillus pentosus* [32], corn stubble has also been used to produce lactic acid by *B. coagulans* [33], coffee mucilage with *L. delbrueckii subsp. bulgaricus* [34], etc.

3.1 Whey

Today, the dairy industry is researching new technologies and looking for new products not only to meet consumer demand through innovative products but also to increase profitability. One of these alternatives is the addition of whey as a substitute for water [35]. However, some laws prohibit the sale of products derived from whey, to prevent food fraud.

Whey contributes with nutrients such as essential amino acids, in addition to the reduction of calories and improvements in technological properties [36]. The addition of this ingredient produces changes in the structure and in the case of chocolate drinks, it produces modifications in the functional and sensory properties [37].

The use of whey as an ingredient in food is regulated, for example, the FDA establishes a maximum of 5%. In white chocolate [38], a similar percentage of 5% of the total mass of chocolate has been established by the EU [39]. The use of *Lactobacillus* in whey as a substrate gives rise to new products, for example, enzymes with antihypertensive properties, which have been isolated after the fermentation of whey with *Lactiplantibacillus plantarum* QS670, *Lactobacillus amylolyticus* L6 has been used to prepare fermented tofu whey, it can be used as a starter culture to produce quagulant tofu and functional drinks [40].

Other studies indicate the suitability in the development of new foods based on lactic serum [41], whose fermentation converts its complex elements into simpler elements, facilitating assimilation in the intestinal tract [42]. Thus, the presence of probiotics increases the number of digestible amino acids, making fermented dairy products a good alternative as a source of nutrients, especially for diseases such as diarrhea [43].

An investigation ensure that salty whey fermented with the microorganism *L. acidophilus* 43S, fights harmful microorganisms such as *Escherichia coli* [6]. In addition, the fermentation of the whey favors the formation of peptides that can improve the functionality of food and beverages by reducing the taste of cheese that the whey has [17]. The way to obtain a probiotic drink based on milk serum is to drain, filter, and submit to a caloric process such as pasteurization, to later inoculate and ferment with the microorganisms of our choice. It is important to mention that the use of whey obtained by curd coagulation has an advantage over acid whey since, with sweet or deproteinized whey, clearer drinks are obtained without the formation of sediments.

3.1.1 Lactic acid and PLA made from whey

In the dairy industry, cheese is made by acidifying milk, which produces the precipitation of proteins known as casein, called curd. In this process, a by-product called whey is obtained, which contains other proteins such as lactalbumin, lactoferrin, and lactoglobulin, additionally, it has lactose, fat, and minerals such as calcium and iron that are present in milk [44].

Around 1/3 of the dairy production is destined for the production of cheese, of which between 80 and 90% of this volume corresponds to whey, which is considered a pollutant too, due to its high biological oxygen demand (BOD) and chemical oxygen demand (COD), currently, the treatment of waste from the food industry is mandatory and necessary to avoid damage to the environment [44].

Investigations have been carried out due to its multiple compounds and its ease of transformation into another useful product, to get the most out of the benefits it can provide. In the polymer industry, whey is used as a precursor for polylactic acid (PLA), a biodegradable material with uses in industry and medicine. Due to the scarcity and environmental regulations, alternatives for materials more compatible with the environment and independent of those of fossil fuels have been developed, this is why biopolymers such as PLA are adjusted to meet these needs in the industry, it is for the demand to elaborate more of these biopolymers has increased over time promoting a new industry worldwide, in the same way, other studies raise similar proposals with matrices other than PLA, focused mainly on starches and proteins [44].

As mentioned, lactic acid has two optical isomers D (−) and L (+), these can be obtained by fermentation using a LAB. However, by means of the chemical method, racemic DL-LA is always obtained [29]. The optical purity of lactic acid is of utmost importance within different industries that seek the combination of D (−) and L (+) polymers to obtain crystalline PLA with good mechanical properties [45]. There are several studies on the production of L (+) lactic acid; however, little research regarding the production of D (−) through fermentation processes were performed. Normally, the production of lactic acid occurs by batch, but this method has the disadvantage of reducing productivity and production due to inhibition by high concentrations of substrate, due to this, different alternatives have been reported in terms of improving productivity of lactic acid [46].

The effect of the different carbon sources for the development of the *L. delbrueckii subsp. delbrueckii* to the production of lactic acid using by-products like whey was evaluated [16]. Through this study, it was possible to conclude that the production of high optical purity D (−) lactic acid is possible by the microorganism, with the use of waste raw material such as molasses and corn liquor, without the need for pre-treatment of these by-products. This represents a low cost to produce acid with high productivity using a fed-batch strategy [16].

The use of whey for the synthesis of polylactic acid, with the use of lactic acid bacteria, such as *L. delbrueckii subsp. delbrueckii*, through fermentation and its use as biodegradable material for food containers, is an alternative that must be studied yet. This proposal has great technological and environmental relevance because the development and characterization of the physical–chemical and mechanical properties of a biodegradable container from a by-product such as whey, will be able to solve contamination problems in the environment and the possibility of showing that natural polymers generate a greater contribution to the term of biodegradability [44].

3.2 Meat by-products

LAB is known to be present in the fermentation processes of meat products and by-products, to produce metabolites such as lactic and acetic acid. To carry out the fermentation of these meat products, it is possible to add an inoculum or simply take advantage of the bacterial flora present in the muscle fibers of the animal [17].

The microorganisms that are commonly found in animal meat and are involved in lactic fermentation are *Pediococcus pentosaceus, Pediococcus acidilacti, Lactiplantibacillus plantarum*. These microorganisms have beneficial effects on meat since there is the production of organic acids that favors the decrease in pH, and development of aroma and flavor, in addition to the partial denaturation of meat proteins, favoring the texture of these fermented products [47].

The decrease in pH helps with the elimination of harmful microorganisms that may be present in meat, especially in the viscera of the animal which are parts where pathogens prevail. The safe pH that the meat must reach is 4.0 to 4.2. However, in the case of fish this can take 48 hours to reach this point that, compared to poultry, their viscera reach the desired pH point between 24 to 36 hours [17].

3.3 Farinaceous by-products

Studies mention that to produce lactic acid it is necessary to use substrates rich in starch, such as starch from wheat, corn, rice, etc. [46]. However, these raw materials require pretreatments so that fermentation can be carried out, which is expensive for the industrial process. For this reason, alternatives have been found that can help in the production of this acid with the use of agro-industrial by-products such as molasses and corn liquor. These substrates must comply with the nutritional requirements of the LAB, and in this way reduce the cost of production [48]. For example, a study carried out to analyze and investigate the production of stereospecific lactic acid from agro-industrial by-products using strains belonging to *Lactobacillus* and *Pediococcus* in combination with enzymatic hydrolysis shown that raw materials, wheat bran (WB), distillery grains (DGS), used brewery grains (BSG) and lupine seeds (LF) (*Lupinus angustifolius*), can be efficiently synthesized [15]. In this investigation, favorable results were obtained for the propagation of LAB and the production of lactic acid. The lowest pH that was reached was with LF after 48 hours of fermentation with the strains of *P. pentosaceus* and *Pediococcus acidilactici*, while with the strain of *Latilactobacillus sakei* the substrate with the best result was WB. Thus, it is possible to conclude that with cellulosic agro-industrial waste using LAB, L-lactic acid can be efficiently synthesized [15].

Cassava (*Manihot esculenta*) is a tuber that has a large number of complex carbohydrates, which can lead to the production of lactic acid. This root is made up of 20% bagasse made up of peel and bark and 80% per tuber. Cassava bagasse is composed of 50 to 60% starch, approximately 34% cellulose, 15% hemicellulose, and 7% lignin [49]. In Colombia, studies have been carried out where the production of lactic acid was analyzed using a cassava starch solution and the use of *Lactobacillus* strains grown from yogurt [50]. A study with two different cassava varieties widely in Ecuador inoculated with *L. delbrueckii subsp. lactis* to produce lactic acid showed the sample with high starch presented highest lactic acid at pH of 5.5 and 150 rpm of stirring [51]. The results obtained in this study show that the production of lactic acid increases with pH with which it works during fermentation if the value exceeds 5.5 there is a growth inhibition of lactic acid producing microorganisms from glucose [52]. The determining factors in this work were the substrate and the pH, it is also important to mention that the used microorganism directly consumes glucose to produce lactic acid; however, in the hydrolysis of starch, reducing sugars were obtained that include maltose, dextrins, and glucose. Regarding the pH and density, it was possible to determine a correlation between higher density values with the presence of lactic acid in the product obtained by fermentation, and consequently, the pH of this will be lower. Finally, it is possible to conclude that at higher pH with stirring, the microorganism works optimally in a fermentation process with hydrolyzed cassava bagasse, obtaining relevant yields. Thus, with the analyzes carried out on the lactic acid obtained, its quality was verified, and it may have value in the market for the sustainable development of Ecuador. However, the starch obtained from cassava bagasse for fermentation should be studied in greater depth, directly in the hydrolysis of the lignocellulosic material present [51].

Most of the biodegradable packaging developed for food use is based on the use of starches, among them those of cassava, potato, and achira, these by-products are subjected to fermentation by LAB to obtain lactic acid, precursor of PLA [51], also proteins such as zein are used, in both cases, the use of a source food to produce an inedible product poses a risk to food safety. There are few by-products used as raw material to produce biodegradable packaging, one of them is starch from cassava husk, there are no studies on the production of dairy packaging from PLA of whey from milk which must be inert and heat resistant [44].

Bakery waste is other product that can be fermented for obtain new foods, usually is used mainly to obtain breadcrumbs and for feed for livestock. It is possible to market it as a food with added value, for this, it is necessary to subject the by-products to crushing and drying to obtain a fine powder with low moisture content for these processes such as drying, mixing with other elements, and crushing is important [17]. Bakery by-products and other food waste have been subjected to fermentation to improve their physicochemical characteristics using LAB such as *Ligilactobacillus salivarius* [53]. The changes that can be seen due to anaerobic fermentation with LAB is the increase in soluble carbohydrates and nutritional improvement. However, it is achieved through a short fermentation of 10 days; thus, the microbial load is not so elevated. To carry out a fermentation process like this, it was considered that 0.2% inoculum is the optimal amount for improving the nutritional properties of bakery coproducts using LAB [47].

3.4 Fruit by-products

To produce lactic acid, the use of fruit waste as liquid pineapple could be beneficial since it is an element rich in glucose and nutrients [14]. The pineapple canning industry is one of the food industries that generates a large amount of both solid and liquid waste and that must follow strict environmental regulations. For this reason, there is a special interest in this productive sector. In addition, the use of effluents as carbon sources for lactic acid fermentation helps to reduce or eliminate pollution and reduce costs [14]. However, there is the presence of metals such as copper, zinc, magnesium, calcium, iron, etc., which can cause problems in the fermentation to produce lactic acid. These can inhibit the growth of LAB, influence the pH of the substrate, and also be involved with the inactivation of enzymes that participate in the synthesis of products [26]. Currently, the most widely used microorganism immobilization matrix has been sodium alginate, this matrix has been used for *Saccharomyces cerevisiae, Bacillus amyloliquefaciens,* and *Kluyveromyces*. The advantage of the use of this matrix is its stability; The substrates and products easily diffuse in and out. However, there is much work on the production of lactic acid, the use of pineapple residues and the immobilization of lactic microorganisms have not been explored [54]. In this study, it was found that sodium alginate in a concentration of 2% generates the maximum production of lactic acid in comparison with the other concentrations tested. Regarding pH, the best yield in the production of lactic acid was 6.5 and finally, the optimal temperature during fermentation was 37°C [14]. Thus, it is possible to conclude that the production of lactic acid using effluents from the pineapple canned industry is viable, if the optimal sodium alginate conditions, the working pH, and the temperature are considered [14].

3.5 Marine by-products

The main elements of interest from shellfish by-products are pigments such as carotenoids and melanin. Carotenoids are responsible for the coloring of the meat and skin of some fish such as salmonids, they have warm colors such as yellow, orange, and red, Furthermore, they are found in the exoskeletons of crustaceans such as crabs, lobsters, and shrimp. Melanin is a dark pigment that ranges from brown to black, this product of an oxidation reaction of phenolic compounds is present in the peritoneal lining, skin, and eyes of some species [17].

For the extraction of carotenoids from the by-products of crustaceans, several studies have been carried out, using enzymatic or fermentative methods, that are more preferable for compounds such as carotenoids, thus, obtaining higher yields and higher quality carotenoids for their possible applications [17]. Regarding very unstable carotenoids, the silage method by fermentation with lactic acid has been tried, in order to stabilize astaxanthin [55]. The extraction of carotenoids by means of silage of shrimp by-products (*L. delbrueckii subsp. indicus*) by *Lactiplantibacillus plantarum* was studied, in addition, extraction with a mixture of hexane and isopropanol as solvents and refined sunflower oil and the effect of each of the procedures to analyze the stability of the carotenoids, obtaining that the fermentation turned out to be better compared to acid silage, in solvent as in oil [17].

4. Conclusion

The use of agro-industrial by-products with LAB, specifically *Lactobacillus* spp., and consequently the production of lactic acid, fits within the ideology and concept of circular economy, focusing on the pillars of economic, ecological, and social aspects. The use of material that is usually discarded represents a relief for the planet in terms of the reduction of polluting matter that thanks to innovation can be used, the reduction of costs for companies by substituting more expensive raw material for another more economically accessible, and the social contribution through the generation of employment and economic activation [51]. This favorable impact, sought by the United Nations with the fulfillment of the Sustainable Development goals proposed for the year 2030, can also be reflected in the valorization of agro-industrial by-products not only to produce lactic acid but also in other applications. Examples of these applications are nutraceutical formulations and pharmaceuticals through the use of fungal co-products [56], food packaging through the use of whey [57], valorization of lard through biotechnological tools employing fermentation in state solid of *Yarrowia lipolytica* and *Lacticaseibacillus paracasei* [58], biorefining of whey from cheese in order to generate high-value products and eliminate environmental contamination by whey [59], revaluation of by-products of the wine industry through the application of LAB strains [60], use of fruit and vegetable by-products for the development of pharmaceutical products [61], etc.

Author details

Ederson Freire-Almeida and Pedro Maldonado-Alvarado*
Department of Food Sciences and Biotechnology, Escuela Politécnica Nacional, Quito, Ecuador

*Address all correspondence to: pedro.maldonado@epn.edu.ec

References

[1] FAO. Agricultural Biotechnologies: Agroindustria [Internet]. Las biotecnologías en la agroindustria en los países en desarrollo. 2010. Available from: https://www.fao.org/biotech/sectoral-overviews/agro-industry/es/. [Accessed: October 25, 2021]

[2] Vargas Corredor YA, Peréz Pérez LI. Aprovechamiento de residuos agroindustriales en el mejoramiento de la calidad del ambiente. Rev Fac Ciencias Básicas. 2018;**V(1)**:59-72

[3] Cerantola N, Ortiz Pinilla MT. Materiales sobre la economía y el consumo circular. La Econ Circ en el Sect Agroaliment; 2018. pp. 8-11; p. 13. Available from: https://www.otroconsumoposible.es/publicacion/economia-circular.pdf

[4] Chiocchetti GM, Jadán-Piedra C, Monedero V, Zúñiga M, Vélez D, Devesa V. Use of lactic acid bacteria and yeasts to reduce exposure to chemical food contaminants and toxicity. Crit Rev Food Sci Nutr [Internet]. 2019;**59**(10):1534-1545. DOI: 10.1080/10408398.2017.1421521

[5] Sánchez ÓJ, Barragán PJ, Serna L. Review of lactobacillus in the food industry and their culture media. Revista Colombiana de Biotecnología. 2019;**21**(2):63-76

[6] Wilson T. Temple NJ. Beverages in Nutrition and Health. Humana Press; 2004. p. 427. Available from: https://link.springer.com/book/10.1007/978-1-59259-415-3

[7] Heredia-Castro PY, Hérnández-Mendoza A, González-Córdova AF, Vallejo-Cordoba B. Bacteriocinas de bacterias ácido lácticas: Mecanismos de acción y actividad antimicrobiana contra patógenos en quesos. Interciencia. 2017;**42**(6):340-346

[8] Foliaco Sandoval DJ, Valdez Barros LM, Gómez Nussbaumer SJ, Ganem Torres YP. Determinación de parámetros cláves en el proceso de escalamiento para la producción de Lactobacillus. Mente Joven. 2019;**7**: 45-58

[9] Zalduegui P. Bioquímica de las fermentaciones. Vol. 50. Madrid, Spain: Monogr la Univ Politécnica Madrid; 1975

[10] Fay DL. Bioquimica De Los Microorganismos. Angewandte Chemie International Edition. 1967;**6**(11):951-952

[11] Farazin A, Mohammadimehr M. Effect of different parameters on the tensile properties of printed Polylactic acid samples by FDM: Experimental design tested with MDs simulation. International Journal of Advanced Manufacturing Technology. 2021;**118**:103-118

[12] Garcia C, Arrázola G, Durango A. Produccion de Ácido Láctico por vía Tecnologica. Temas Agrar. 2010;**15**(2):9-26

[13] Romo-Buchelly J, Rodríguez-Torres M, Orozco-Sánchez F. Biotechnological valorization of agro industrial and household wastes for lactic acid production. Revista Colombiana de Biotecnología. 2019;**21**(1):113-127

[14] Idris A, Suzana W. Effect of sodium alginate concentration, bead diameter, initial pH and temperature on lactic acid production from pineapple waste using immobilized lactobacillus delbrueckii. Process Biochemistry. 2006;**41**(5):1117-1123

[15] Juodeikiene G, Klupsaite D, Zadeike D, Cizeikiene D, Vidziunaite I, Bartkiene E, et al. Bioconversion of agro-industrial by-products to lactic acid using lactobacillus sakei and two Pediococcus spp. strains. International Journal of Food Science and Technology. 2016;**51**(12):2682-2691

[16] Beitel SM, Coelho LF, Contiero J. Efficient conversion of Agroindustrial waste into D(−) lactic acid by lactobacillus delbrueckii using fed-batch fermentation. BioMed Research International. 2020;**2020**:13. DOI: 10.1155/2020/4194052

[17] Chandrasekaran M. Valorization of Food Processing by-Products. Vol. 3. CRC Press; 2013. p. 800. Available from: https://www.taylorfrancis.com/books/mono/10.1201/b12816/valorization-food-processing-products-chandrasekaran

[18] Burgo Bencomo OB, Gaitán Suazo V, Yanez Sarmiento J, Zambrano Morales ÁA, Castellanos Pallerols G, Estrada Hernández JA. La Economía circular una alternativa sostenible para el desarrollo de la agricultura. Rev Espac. 2019;**40**(13):1-5

[19] Kowszyk Y, Maher R. Estudios de caso sobre modelos de Economía Circular e integración de los Objetivos de Desarrollo Sostenible en estrategias empresariales en la UE y ALC. Fund EU-LAC. 2018:8-9. Available from: https://eulacfoundation.org/es/system/files/economia_circular_ods.pdf

[20] Cabeza MD. El sistema agroalimentario globalizado: imperios alimentarios y degradación social y ecologica. Rev Econ Crítica. 2010;**10**:32-61

[21] Food Drink Europe. VISIÓN DE SOSTENIBILIDAD MEDIOAMBIENTAL PARA. Vol. 2030. Food Drink Europe; 2012. p. 92. Available from: https://www.yumpu.com/en/document/read/46865820/fooddrinkeurope-environmental-sustainability-vision-towards-2030

[22] Okano K, Tanaka T, Ogino C, Fukuda H, Kondo A. Biotechnological production of enantiomeric pure lactic acid from renewable resources: Recent achievements, perspectives, and limits. Applied Microbiology and Biotechnology. 2010;**85**(3):413-423

[23] John RP, Nampoothiri KM, Pandey A. Fermentative production of lactic acid from biomass: An overview on process developments and future perspectives. Applied Microbiology and Biotechnology. 2007;**74**(3):524-534

[24] Tang Y, Zhao D, Cristhian C, Jiang J. Simultaneous saccharification and cofermentation of lignocellulosic residues from commercial furfural production and corn kernels using different nutrient media. Biotechnology for Biofuels. 2011;**4**(1):1-10

[25] Burgos-Rubio CN, Okos MR, Wankat PC. Kinetic study of the conversion of different substrates to lactic acid using lactobacillus bulgaricus. Biotechnology Progress. 2000;**16**(3):305-314

[26] Kotzamanidis C, Roukas T, Skaracis G. Optimization of lactic acid production from beet molasses by lactobacillus delbrueckii NCIMB 8130. World Journal of Microbiology and Biotechnology. 2002;**18**(5):441-448

[27] Naveena BJ, Altaf M, Bhadrayya K, Madhavendra SS, Reddy G. Direct fermentation of starch to L(+) lactic acid in SSF by lactobacillus amylophilus GV6 using wheat bran as support and substrate: Medium optimization using RSM. Process Biochemistry. 2005;**40**:681-690

[28] Hadadji M, Bensoltane A. Growth and lactic acid production by *Bifidobacterium longum* and *lactobacillus acidophilus* in goat's milk. African J Biotechnol. 2016;**5**(6):505-509

[29] Hofvendahl K, Hahn-Hägerdal B. Factors affecting the fermentative lactic acid production from renewable resources. Enzyme and Microbial Technology. 2000;**26**(2-4):87-107

[30] Yang X, Zhu M, Huang X, Lin CSK, Wang J, Li S. Valorisation of mixed bakery waste in non-sterilized fermentation for l-lactic acid production by an evolved Thermoanaerobacterium sp. strain. Bioresour Technol [Internet]. 2015;**198**:47-54. DOI: 10.1016/j.biortech.2015.08.108

[31] Rojas AM, Montaño LP, Bastidas MJ. Produção de ácido láctico de soro de leite coalhado usando Lactobacillus delbrueckii subsp. Bulgaricus e Streptococcus thermophilus. Revista Colombiana de Quimica. 2015;**44**(3):5-10

[32] Wischral D. Statistical optimization of lactic acid production by lactobacillus pentosus using Hemicellulosic hydrolysate from sugarcane bagasse. Rev Ing. 2018;**29**(1):41-51

[33] Caballero Salinas JC, Moreno Reséndez A, Reyes Carrillo JL, García Valdez JS, López Báez W, Jiménez Trujillo JA. Competencia del uso del rastrojo de maíz en sistemas agropecuarios mixtos en Chiapas. Rev Mex Ciencias Agrícolas. 2017;**8**(1):91-104

[34] Arias Zabala M, Henao Navarrete L, Castrillón GY. Lactic acid production by fermentation of coffee mucilage with lactobacillus bulgaricus NRRL-B548. Dyna. 2009;**76**(158):147-153

[35] Patel S, Prasanth S, Choudhary PL, Sahu C. Techno-economic feasibility of whey based mango-herbal (ginger) beverage. Indian journal of dairy science. 2007;**60**:149-155

[36] Carter BG, Foegeding EA, Drake MA. Invited review: Astringency in whey protein beverages. Journal of Dairy Science. 2020;**103**(7):5793-5804

[37] Lapčíková B, Lapčík L, Salek R, Valenta T, Lorencová E, Vašina M. Physical characterization of the milk chocolate using whey powder. LWT. 2022;**15**(154):112669

[38] Center for Devices and Radiological Health. Medical Device Databases-Code of Federal Regulations-Title 21- Food and Drugs. Center for Devices and Radiological Health. 2018. Available from: https://www.ecfr.gov/current/title-21

[39] WHO. Standard for Chocolate and Chocolate Products. WHO; 2016. Available from: https://www.fao.org/fao-who-codexalimentarius/sh-proxy/en/?lnk=1&url=https%253A%252F%252Fworkspace.fao.org%252Fsites%252Fcodex%252FStandards%252FCXS%2B87-1981%252FCXS_087e.pdf

[40] Fei Y, Liu L, Liu D, Chen L, Tan B, Fu L, et al. Investigation on the safety of lactobacillus amylolyticus L6 and its fermentation properties of tofu whey. LWT-Food Sci Technol. 2017;**84**:314-322

[41] Djurić M, Carić M, Milanović S, Tekić M, Panić M. Development of whey-based beverages. European Food Research and Technology. 2004;**219**(4):321-328

[42] Khedkar JN, Choudhari DM, Pawar BK, Kadam VS. Use of milk by-product (whey) for preparation of beverages. Asian J Anim Sci. 2014;**9**(2):193-197

[43] Hitchins AD, FE MD. Prophylactic and therapeutic aspects of fermented milk. The American Journal of Clinical Nutrition. 1989;**49**(4):675-684

[44] Zapata A, Pujol O, Coda F. Biodegradable polymers: An alternative of future to the sustainability of the environment. Técnica Ind. 2012;**297**:76-80

[45] Lunt J. Large-scale production, properties and commercial applications of poly lactic acid polymers. Polymer Degradation and Stability. 1998;**59**(1-3):145-152

[46] Hujanen M, Linko S, Linko YY, Leisola M. Optimisation of media and cultivation conditions for L(+)(S)-lactic acid production by lactobacillus casei NRRL B-441. Applied Microbiology and Biotechnology. 2001;**56**(1-2):126-130

[47] Toldrá F, Hui YH, Astiasarán I, Nip WK, Sebranek JG, Silveira ETF, et al. Handbook of Fermented Meat and Poultry. Wiley; 2008. pp. 1-555. Available from: https://onlinelibrary.wiley.com/doi/book/10.1002/9781118522653

[48] Busairi AM. Effect of nitrogen source and initial sugar concentration on lactic acid fermentation of pineapple waste using *L. delbrueckii*. TEKNIK. 2010;**31**(1):31-34

[49] Sivamani S, Baskar R. Process design and optimization of bioethanol production from cassava bagasse using statistical design and genetic algorithm. Preparative Biochemistry & Biotechnology. 2018;**48**(9):834-841. DOI: 10.1080/10826068.2018.1514512

[50] Ortíz C, Ospina L, Sánchez S. Producción de ácido Láctico a partir de Almidón de Yuca por bacterias acidolácticas del género Lactobacillus. Bogotá, Colombia: Univ Bogotá Jorge Tadeo Lozano; 2014

[51] Inguillay S, Jadán F, Maldonado-Alvarado P. Fermentation study of cassava bagasse starch hydrolyzed's using INIAP 650 and INIAP 651 varieties and a strain of lactobacillus leichmannii for the lactic acid production. Rev Bionatura. 2021;**6**(2):1803-1811

[52] Pandey A, Soccol C, Nigam PS-N, Soccol V, Vandenberge L, Mohan R. Biotechnological potential of agro-industrial residues. II: Cassava bagasse. Bioresource Technology. 2000;**74**(1):81-87

[53] Yang SY, Ji KS, Baik YH, Kwak WS, McCaskey TA. Lactic acid fermentation of food waste for swine feed. Bioresource Technology. 2006;**97**(15):1858-1864

[54] Abdel-Naby MA, El-Refai HA, Abdel-Fattah AF. Biosynthesis of cyclodextrin glucosyltransferase by the free and immobilized cells of Bacillus cereus NRC7 in batch and continuous cultures. Journal of Applied Microbiology. 2011;**111**(5):1129-1137

[55] Bhaskar N, Suresh PV, Sakhare PZ, Sachindra NM. Shrimp biowaste fermentation with Pediococcus acidolactici CFR2182: Optimization of fermentation conditions by response surface methodology and effect of optimized conditions on deproteination/demineralization and carotenoid recovery. Enzyme and Microbial Technology. 2007;**5**(40):1427-1434

[56] Antunes F, Marçal S, Taofiq O, Morais AMMB, Freitas AC, Ferreira ICFR, et al. Valorization of mushroom by-products as a source of value-added compounds and potential applications. Molecules. 2020;**1**(25):1-40

[57] Rollini M, Musatti A, Cavicchioli D, Bussini D, Farris S, Rovera C, et al. From cheese whey permeate to Sakacin-a/bacterial cellulose nanocrystal conjugates for antimicrobial food packaging applications: A circular economy case study. Scientific Reports.

2020;**10**(1):1-14. DOI: 10.1038/s41598-020-78430-y

[58] Cotârleţ M, Stănciuc N, Bahrim GE. Yarrowia lipolytica and lactobacillus paracasei solid state fermentation as a valuable biotechnological tool for the pork lard and okara's biotransformation. Microorganisms. 2020;**8**(8):1-12

[59] Lappa IK, Papadaki A, Kachrimanidou V. Cheese whey processing: Integrated biorefinery. Food. 2019;**8**:1-37

[60] Hernández-Macias S, Comas-Basté O, Jofré A, Bover-Cid S, Latorre-Moratalla ML, Vidal-Carou MC. Growth-promoting effect of cava lees on lactic acid bacteria strains: A potential revalorization strategy of a winery by-product. Food. 2021;**10**(7):1-10

[61] Brito TBN, Ferreira MSL, Fai AEC. Utilization of agricultural by-products: Bioactive properties and technological applications. Food Review International. 2020;**00**(00):1-25. DOI: 10.1080/87559129.2020.1804930

Chapter 6

Lactobacilli: Application in Food Industry

Nagina Rafique, Turfa Mamoona, Saiqa Bashir, Imtiaz Hussain and Imran Hayat

Abstract

Lactobacillus is an important class of Gram-positive, non-spore-forming bacteria for food industrial applications. The genus *Lactobacillus* is a potential candidate in fermentation technology for the production of fermented food, feed, and pharmaceutical products. The diverse features of *Lactobacilli* based on their capability to produce acids, enzymes, bacteriocins by fermenting carbohydrates. *Lactobacilli* have probiotic potential and therefore applied in dairy [cheese, yoghurt, fermented milk] and nondairy products such as sausages, juices as well as in animal feed in the form of starter culture. Among *Lactobacilli*, lactic acid-producing bacteria are applied as starter cultures in a variety of fermented foods. *Lactobacilli* are the natural microflora of the gastrointestinal tract and play a beneficial role against infections. The ability of *Lactobacilli* to produce bacteriocins and other antifungal compound lead to the development of bioprotective cultures for use in different foods. Bacteriocins has wide applications in food industries for preventing the attack of foodborne pathogens and for manufacturing active packaging materials. This chapter aimed to review significant industrial applications of *Lactobacilli* with specified strains and also starter cultures with their potential beneficial effects are engrossed. The chapter highlights the commercial applications of *Lactobacilli* in the food, feed, wine and pharmaceutical industries.

Keywords: *lactobacillus*, zoo-technical, probiotics, *Lb. rhamnosus*, Bacteriocins

1. Introduction

The genus *Lactobacillus* was discussed for the first time in 1901. These are rod-shaped, nonspore-forming, Gram-positive bacteria found in fermented foods and are member of the gastro-intestinal tract and vaginal cavity of humans [1–4]. *Lactobacillus* is a group of bacteria with diverse species having higher GC content ranging from 32 to 53 mol% [3]. Morpho-microscopically *Lactobacilli* are rod shaped, non-motile and varying in length ranging from short to long. *Lactobacilli* are comprised a group of multifunctional microbes with shelf life extending characters of food as well as improving human health in the form of probiotics [4]. Studies of the whole bacterial genomic sequence enabled, the scientists to reclassify the species of *Lactobacilli* into 25 genera including an edited genus of *Lactobacillus* i.e. *Paralactobacillus* and *Lactobacillus delbrueckii* group and 23 other novel genera containing *Acetilactobacillus, Agrilactobacillus, Amylolactobacillus, Apilactobacillus, Bombilactobacillus,*

Companilactobacillus, Dellaglioa, Fructilactobacillus, Furfurilactobacillus, Holzapfelia, Lac ticaseibacillus, Lactiplantibacillus, Lapidilactobacillus, Latilactobacillus, Lentilactobacillus, Levilactobacillus, Ligilactobacillus, Limosilactobacillus, Liquorilactobacillus, Loigolactobacilus, Paucilactobacillus, Schleiferilactobacillus, and *Secundilactobacillus* [5]. Overall, more than 250 species have been assigned to the genus *Lactobacillus previously and* are now referred to with a new genus name. *Lactobacillus* releases lactic acid as the major end product of metabolism during fermentation of glucose with small amounts of acetic and succinic acid [2, 3].

The most common habitat of *Lactobacilli* is the gastro-intestinal tract of mammals where these bacteria may colonize. This group of bacteria is also commonly found in mouth such as in tooth plaque, saliva, vaginal tract of humans and mammals [2, 3]. *Lactobacillus* species commonly isolated from GI tract are *Lactobacillus acidophilus, L. brevis, L. plantarum, Lactobacillus salivarius and Lactobacillus fermentum*. *Lactobacillus* are significantly required for gastrointestinal health and are not considered as a pathogen in healthy individuals except when having an association with dental caries [6, 7]. *Lactobacilli* are usually reflected as proactive microbes as they inhibit the growth of pathogenic microorganisms by producing lactic acid and other metabolic compounds [8]. While in immunocompromised patients' lactobacilli are intricate as pathogens and various reports of AIDS, organ transplant, neutropenia caused by *lactobacillus* infection has been published [9–11]. The most common infections caused by *Lactobacilli* are abscesses, bacteremia, endocarditis, neonatal meningitis, dental caries and chorioamnionitis [5, 12]. The chapter focuses on an exploration of the industrial potential of *Lactobacilli* for manufacturing a diversity of products. The purpose of this chapter is to gather literature related to the utilization of *Lactobacillus* species as a microbial candidate for commercial-scale production of valuable products for improving human and animal health by protecting them against pathogenic microflora. The major emphasis is to strive for the attention towards the sustainable application of potent *Lactobacillus* species for futuristic prosperities of mankind.

2. *Lactobacillus* in food industry

Lactobacillus has a long history of use in biotechnology predominantly in manufacturing and conserving food ingredients by the process of fermentation, yoghurt, cheese, kefir, sauerkraut and many other fermented food products productions with the addition of animal feeds manufacture [13]. Further bioprocess technology has developed the specified production aids and procedures for food and feed ingredients [14].

Lactobacilli were the first microbes used by human beings for processing foods and preservation of foods by inhibiting other microbial invasions that can cause food spoilage and ultimately results in foodborne illness [15, 16]. *Lactobacillus* is a significant genus to recent food and feed technologies, not only because of its cumulative interest in valuable functional properties. The dairy industry and self-care health productions are dynamically engaged in promoting the usage of ***Lactobacilli*** in food, [17]. Some traditional fermented foods (**Table 1**) based on ***Lactobacillus*** fermentation from different origins, cultures, civilizations, customs, social relations and sometimes customized with religions were reported by Food and Drug Administration [18, 19]. Because of the characteristic to produce organic acids by using carbohydrates ***Lactobacilli*** have extensive application in the food industry [20].

Name	Country of origin	Description	*Lactobacillus* Species
Gundruk	Nepal	Prepared by fermentation of leafy vegetables. It is an appetizer.	*Lb. cellobiosus* and *Lb. plantarum*.
Kefir	Caucasus	Fizzy fermented milk	*Lb. kefir*, *Lb. casei*, *Lb. acidophilus*, *Lb. bulgaricus*
Coffee	Africa, Asia and Latin America	A fine dark brown powder made from roasted coffee beans. Brewed with boiling water and consumed as a drink	*Lb. plantarum*; *Lb. brevis*.
Kishk	Egypt	Dried mixture of fermented milk with cereal.	*Lb. bulgaricus*, *Lb. plantarum*, *Lb. casei*, *Lb. brevis*
Laban zeer	Egypt	Concentrated sour buttermilk used in the manufacturing of Kishk, sometimes used in salad or to make a beverage after dilution with water; semisolid consistency, tart and salty taste	*Lb. casei*, *Lb. plantarum*, *Lb. brevis*
Koumiss	Kazakhstan and Kyrgyzstan	Milk is processed into a drink by lactic acid and alcoholic fermentation	*Lb. salivarius*, *Lb. buchneri*, *Lb. helveticus*, *Lb. plantarum*, *Lb. acidophilus*
Zabaday	Egypt	Set fermented milk; characteristic taste and aroma; full, pleasant, mildly sour taste	*Lb. bulgaricus*, *Lb. casei*, *Lb. fermentum*, *Lb. helveticus*, *Lb. viridescens*.
Zincica	Czech Republic	Salted kefir-like beverage made from coagulated whey proteins	*Lb. casei*, *Lb. plantarum*, *Lb. lactis*

Table 1.
Traditional foods containing lactobacilli.

2.1 Lactobacilli as starters in food processing industries

Lactobacilli are mainly used as a starter culture in the contemporary food processing industry. The starter culture may contain only one pure strain or a combination of strains from different species of microbes [20]. Starter cultures can be well-defined as the enormous number of cell preparations, any of a single strain type or an assortment of two or more microbial strain/species that are supplemented in foods to get the advantage of consequent products or compound obtained from their metabolic or enzymatic activity [21]. Since starter cultures are used in food production settings to accomplish processes of fermentation, their usage is globally a common practice in food manufacturing industry. This has led to the commercialization of numerous products such as probiotics or starters, bio-protective cultures that are intended to deliver foods with specific nutritional and sensual characteristics, prospective health benefits and assurance of food safety [22]. Starter cultures are practically applied in a wide array of food industries such as the dairy productions for the manufacture of yoghurt, cheese and other fermented

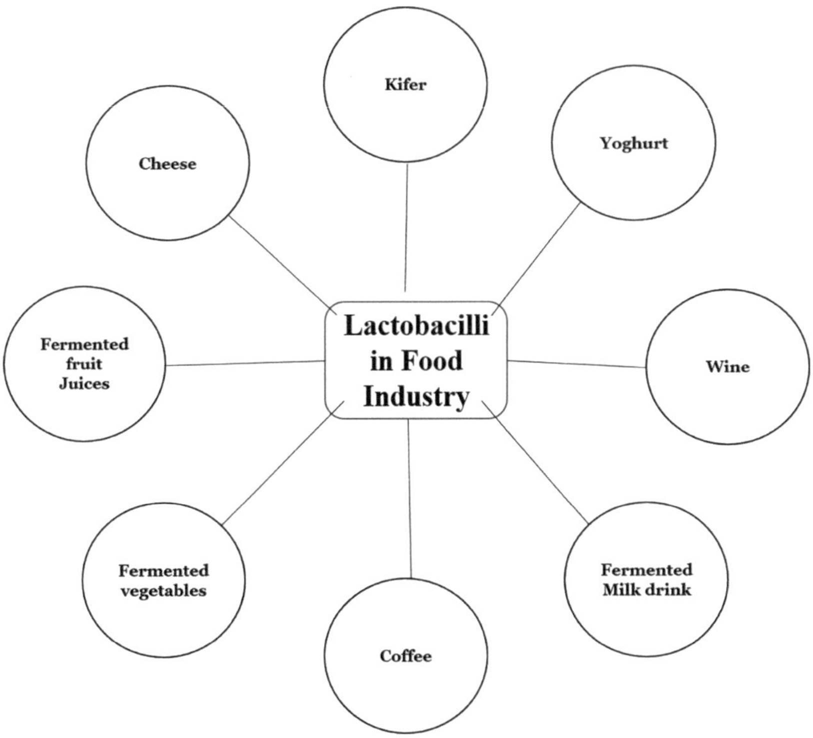

Figure 1.
Application of lactobacillus in food industry.

products from dairy milk [23], sausage manufacture in meat industry [24], in wine and beer industry for the production of alcohol [25, 26], production of vinegar [27], preparation of soy and rice-based oriental foodstuffs [28], fermented cereals and bakery products [29], and in making of fermented products from fruits and vegetables [30–32]. **Figure 1** clearly illustrates the industrial potential of lactobacilli.

Starter cultures may contain any type of microbes such as bacteria, yeast, mold or any type of fungal culture. But the species of *Lactobacilli* reported in the literature have vast potential for usage as starter culture because of its antimicrobial effects which ultimately prove beneficial for food safety. Potential antimicrobial properties of some lactobacilli have been described by scientists such as *Latilactobacillus sakei* isolated from meat sausages have an antimicrobial effect against *Listeria monocytogenes*, *Salmonella spp.* and *E. coli* in combination of garlic powder and wine [33]. Another strain of *lactobacillus* i.e. *Lactiplantibacillus plantarum* was found effective against pathogenic bacteria of the *Enterobacteriaceae* family when used as a starter in Vacuum packaged products [34].

2.2 Lactobacilli in dairy industry

The *Lactobacillus* genus comprise a large number of diverse species with relatively large degree of variation and diversity. It is one of the largest genus in the group of lactic acid bacteria of over fifty species [35]. *Lactobacilli* are the important lactic acid producing bacteria and are responsible for the development of microflora in the most dairy products particularly cheese and fermented milk. *Lactobacillus* is the significant microorganism for development of color, flavor and texture of dairy products via acidification owed to lactic acid and by metabolizing milk proteins. The commonly used species of L**actobacilli** in the dairy industry are *Lb. lactis*, *Lb. casei*, *Lb. rhamnosus*, *Lb. helveticus*, *Lb. plantarum* and *Lb. curvatus* [36]. Moreover, *lactobacilli* are incorporated in the preparation of cheese, yoghurt and fermented milk as probiotic cultures because of their potential benefits for the treatment of acute and chronic inflammation of the gastrointestinal tract [37]. *Lactobacilli* can additionally be used for the preservation of dairy products as they can produce bacteriocins which are usually inhibiting the growth of other non-desired microbes [38]. The species of *Lb. delbrueckii* encompasses three subspecies that is *subsp. bulgaricus*, subsp. *delbrueckii* and subsp. *lactis* [35] are used as a starter culture for yoghurt preparation. The information related to the use of other cultures for yoghurt manufacturing and safety is scarce.

The strain *Lb. paracasei subsp. paracasei* is recovered commonly from ripened cheese and forms a constitute with *Lb. curvatus*, *Lb. plantarum*, *Lb. casei* and *Lb. rhamnosus* is the primary microbiota of the non-starter Lactic acid bacteria group contributing to the process of cheese maturation [39, 40]. Among the species of lactobacilli *Lb. rhamnosus* is one that has been applied as a probiotic organism in functional foods. The strain *Lb. rhamnosus* has been identified and designated with a number HN001, has both probiotic and flavor improving qualities, hence, it can be useful as an adjunct in the manufacturing of cheese to decrease dangerous microflora, improve the flavor of cheese and increase the speed of cheese ripening [41, 42]. *Lactobacillus johnsonii* formerly known as *Lb. acidophilus* has been widely studied because of its probiotic characteristic and is commercialized for production of fermented milk (LC1) products [43]. The *Lb. johnsonii* displays antimicrobial [44–46] and immunomodulatory properties [47, 48] properties. A thermophilic starter culture of *Lactobacillus helveticus* is used in the production of a numeral fermented products of dairy [49] and develops on a comparatively limited number of carbohydrates such as galactose and lactose with typical requirement of few vitamins for growth [50, 51]. Knowledge about the biodiversity of ***Lactobacillus*** can be employed as a starter culture in different dairy products described in previous literature is provided in the tabulated form (**Table 2**).

2.3 Lactobacilli in non-dairy food products of plant and animal origin

The species of *Lactobacillus* are useful for production of fermented foods other than dairy such as sauerkraut by fermenting cabbage. Usually, normal microflora of cabbage with added *Lb. plantarum* act as starter and salt like sodium chloride is used for preventing the growth of foodborne pathogenic microorganisms [68]. An additional study reported the use of both *Lb. plantarum* and *Lb. brevis* for production of sauerkraut [69]. Another food product in which the genus *lactobacillus* is involved is pickle where the propagation of starter culture of *Lb. plantarum* is responsible for achieving the desired effect. Another investigation reported the involvement of *Lb. versmoldensis* sp. nov.; *Lb. plantarum*; *Lb. casei*; *Lb. pentosus* for the production of fermented green olives [70–72]. The ***Lactobacillus*** species such as *Lb. sakei* and *Lb. curvatus* were found in dry sausages [73–75]. A previous study reported the presence of *Lb. sanfranciscensis*, *Lb. reuteri* and *Lb. pontis* in sourdough [76].

L. acidophilus was reported in chocolate by Rosell [77] and *Lb. plantarum* was found in juice drinks [78] as well as *Lb. rhamnosus* was described as a microbe found in juice drinks (Valio Gelfilus) GG (ATCC 53103) [64].

Products	Diversity of Lactobacilli
Unpasteurized milk	*Lb. casei subsp. casei*; *Lb. paracasei subsp. paracasei*; *Lb. rhamnosus*; *Lb. plantarum*; *Lb. fermentum*; *Lb. brevis*; *Lb. buchneri*; *Lb. curvatus*; *Lb. acidophilus*; *Lb. pentosus* [52]
Cheese	*Lb. helveticus*; *Lb. delbrueckii subsp. lactis*; *Lb. delbrueckii subsp. bulgaricus*; *Lb. casei*; *Lb. paracasei subsp. paracasei*; *Lb. paracasei subsp. tolerans*; *Lb. rhamnosus* [53]
Natural Argentine fermented hard cheese	*Lb. delbruecki subsp. lactis*; *Lb. helveticus* [54]
Fioro Sardo	*Lb. paracasei*; *Lb. plantarum*; *Lb. rhamnosus*; *Lb. pentosus*; *Lb. paraplantarum*; *Lb. sake*; *Lb. graminis* and *Lb. curvatus* [55]
Emmental, Comte	*Lb. casei*; *Lb. helveticus*; *Lb. delbrueckii ′ subsp. lactis* [56]
Camembert	*Lb. paracasei subsp. paracasei*; *Lb. plantarum*; *Lb. delbrueckii subsp. lactis*; *Lb. acidophilus*; *Lb. delbrueckii subsp. bulgaricus*; *Lb. casei subsp. casei* [57]
Batzos, a traditional Greek cheese from unpasteurized goat's milk	Lb. plantarum; Lb. paraplantarum; Lb. paracasei subsp. tolerans; Lb. sake; Lb. curvatus; Lb. pentosus [58]
Mozzarella	*Lb. delbrueckii subsp. bulgaricus*; *Lb. plantarum* [59]
Ricotta	*Lb. acetotolerans*; *Lb. alimentarius*; *Lb. gasseri*; *Lb. hilgardii*; *Lb. zeae*; *Lb. brevis*; *Lb. plantarum*; *Lb. paraplantarum* [60]
Yoghurt	*Lb. delbruecki subsp. bulgaricus* according to the Codex Alimentarius standard [61]
Fermented milk	*Lb. kefir* [62]
Kumys	*Lactobacillus delbrueckii subsp. bulgaricus* [63] Adapted from Codex Alimentarius (2011)
Infant formula	*Lb. rhamnosus* GG (ATCC 53103) http://www.valio.f [64]
Frozen yoghurt	*Lb. delbrueckii subsp. bulgaricus*; *Lb. acidophilus* [65]
Ice cream	*Lb. acidophilus*; *Lb. johnsonii* La1 [43, 66]
Fruit yoghurt	*Lb. casei* [67]
Probiotic yoghurt drink	*Lb. acidophilus* La5 and *Lb. casei*
Stirred yoghurt	*Lb. lc1*
Fruit layer yoghurt	*Lb. acidophilus* La5 *LBifidobacterium animalis subsp. lactis* BB12
Fermented soya drink	*Lb. acidophilus*
Yoghurt drink	*Lb. acidophilus* La5
Fromage frais blanc	*Bifidobacterium*, *Lb. acidophilus*
Live natural yoghurt	*Lb. acidophilus*, *Lb. casei*,
Yoghurt drink (Yakult)	*Lb. casei*
Fruit yoghurt smoothie	Yoghurt culture, *Lb. acidophilus*, *Bifidobacterium*
Goat milk yoghurt	*Lb. acidophilus*, *Lb. bulgaricus*, *S. thermophilus*, *Bifidobacterium*
Fermented milk drink	*Lb. casei* Shirota
Fruit yoghurt drink	*Lb. casei*
Bio pouring yoghurt	Probiotic
Fat-free yoghurt drink	*Lb. casei*
Natural Greek style yoghurt	*Lb. acidophilus*, *Lb. bulgaricus*, *S. thermophilus*

Products	Diversity of Lactobacilli
Goat fruit yoghurt	*S. thermophilus, Lb. casei*
Natural goat yoghurt	*Bifidobacterium longum, Lb. acidophilus*
Fruit yoghurt	*B. animalis subsp. lactis, Lb. acidophilus*
Thick and creamy yoghurt	*Bifidobacterium, Lb. acidophilus, S. thermophilus*
Natural fresh and mild yoghurt	*Lb. acidophilus, B. longum, S. thermophilus*
Organic natural yoghurt	*Lb. acidophilus, Bifidobacterium*

Table 2.
Species of lactobacilli present naturally or added as starter in dairy products.

L. fermentum has been used extensively as a probiotic strain and may be found in cereal-based fermented foods and many other vegetables [79]. The predominant bacteria of sourdough and cereal based fermented products is *Lb. plantarum* and is a leading microbe because it applied for producing corn dextrins after the reduction of fermentable sugars [80] *Lactobacillus sanfranciscensis* is the leading lactic acid bacteria in sourdough [80] and 14 genome assemblies of *Lb. sanfranciscensis* have been reported [35].

A fermented beverage tea known as Kombucha is prepared by the traditional fermentation process of fermenting sweetened black tea with tea fungus which contains a consortium of yeasts and acetic acid-producing bacteria. The viability of selected strains of *Lactobacilli* throughout the fermentation process of Kombucha and their interaction with tea fermenting fungus and their role in obtaining a tea beverage with improved functional properties were tested by scientists [81] Five wild strains (*Lactobacillus hilgardii, L. fermentum*, three strains of *L. plantarum*) isolated from conventional fermented foods and were added separately in Kombucha on the second day of fermentation. The addition of wild species of *Lactobacillus* during fermentation of Kombucha contributed significantly to increasing the lactic acid content of the tea beverage. The strain of *Lactobacillus plantarum* was reported for the highest lactic acid production during Kombucha fermentation. The *L. hilgardii* strain isolated from sour dough exhibited the highest sustainability in Kombucha and also has possible probiotic characteristics. So this strain can be used in Kombucha fermentation.

2.4 Lactobacilli in wine industry

The major role of lactic acid bacteria in wine production is to conduct the process of **malolactic** fermentation. The process of wine fermentation reduces the acidity in wine, enhances the mouth feel and aroma and also improves the microbial stability of wine. **Therefore,** *Lactobacilli* **have application in** the **production process of wine from both grapes and fruits** cider [82]. During the process of malolactic fermentation malic acid is converted into lactic acid and carbon dioxide which is extensively required for the maturation of fruit wine [83]. The recent efforts focused towards discovering the biodiversity of geographic areas associated for wine production with the purpose of finding new strains of lactic acid bacteria which can be used as starters [84]. For instance, two potent new autochthonous malolactic fermentation starters i.e. *Lactobacillus paracasei* UVI-2 and *Lb. hilgardii* UVI-23, have been isolated and identified from Albariño grapes in Val do Salnés, Spain [84]. The following species of ***Lactobacilli*** such as *Lb. plantarum* [85], *Lb. brevis*; *Lb. collinoides*; *Lb. hilgardii*; *Lb. paracasei*; *Lb. pentosus*; *Lb. plantarum*; *Lb. mali* [86] are naturally present or added as a starter during the fermentation processing of wine. *Lactobacillus sakei subsp. sakei* was first isolated from rice alcoholic beverage i.e. sake and it is the product from which its name was taken [35].

3. Lactobacilli in feed industry

Many potential strains of ***Lactobacilli*** having probiotic effect are applied in feed production for poultry and other animals for beneficial purposes. There is no case of foodborne illness has been reported by ingestion of feed containing ***Lactobacilli*** [12]. In agricultural practice, ***Lactobacilli*** have long been applied for the preservation of grass or maize in the form of fodder [87] where these bacteria accelerate the rate of decline in pH while protecting plant carbohydrates through homofermentation and also conserving proteins from plants by reducing proteolysis and by deamination [88–91]. According to literature the species of ***Lactobacillus*** **Lb. casei**, *Lb. buchneri* [92], *Lb. paracasei subsp. paracasei* [93], *Lb. plantarum* [94], *Lb. buchneri*, *Lb. fermentum* [95] and *Lb. diolivorans* sp. nov. [96] have been applied in silage production. Probiotic ***Lactobacilli*** are used as an alternative to antibiotics in animal feed because of the safety concerns of antibiotics [97]. Probiotic microbes in animal feed should be live and have zootechnical properties to convert feed and increase weight gain [98]. The supplementation of probiotics has been suggested for handling numerous stress conditions or for preventing diseases in several animal species [17]. The studies which have highlighted the value of lactobacilli

Animal species	*Lactobacillus* species	Benefits
Poultry industry	*Lactobacillus johnsonii* FI9785	Controls endemic necrotic enteritis due to *Clostridium perfringens*, reducing economic losses associated with the decrease in antibiotic use [99]
Chicks	*Lb. acidophilus*	Increases body weight gain, decreases fecal weight [100]
Broiler chickens	*Lb. acidophilus*, *Lb. casei*	Enhances production performance [101]
Late laying period hens	*Lb. species*	Increases egg production, decreases mortality, improves feed conversion but not egg quality [102]
Beef cattle	*Lb. acidophilus*	Reduces *E. coli* O157 levels [103]
Veal calves	Six *Lactobacillus sp.*	Enhances growth rate, body weight gain, Feed conversion and tends to decrease mortality [104]
Dairy cows	*Lb. acidophilus*, *Lb. buchneri* 40,788	Significant increase (16.2%) in milk production Used to treat alfalfa silage. Increases the concentration of acetic acid. When fed to lactating cows, it improved the aerobic stability of the ration and increased milk production [105, 106]
Lambs	*Lb.* species *Lb. acidophilus*, *S. faecium*, *Lb.casei*, *Lb. fermentum* and *Lb. plantarum*	Stimulates feed intake and daily weight gain Reduces fecal shedding of *E. coli* O157:H7 [107, 108]

Table 3.
Health promoting properties of lactobacilli for animal.

for various animal species are tabularized herein (**Table 3**). Probiotic treatment is becoming progressively popular in veterinary medicine, particularly for pets.

Some feed additives of European origin such as BIACTONs N1 12 for piglets contain *Lb. farciminis* [109]; PROVITA LEs N1 E1706 for calves & piglets contain *Lb. rhamnosus* DSM 7133 and *Enterococcus faecium* DSM 7134 [109]; CECT4 529 N123 feed for laying hens hold *Lb. acidophilus* [110] and *Lb. acidophilus* was added to the cats and dogs feed [111]. Other Non-European feed additives such as Silage (PRO-STOREs USA) have *Lb. casei*; *Lb. plantarum*; *Lb. brevis* and other lactic acid bacteria and fungi [112]. HP BIOs a feed for cows in USA comprised of *Lb. fermentum*; *Lb. acidophilus*; *Lb. plantarum*; *Lb. casei*; and lactic acid bacteria [112]. Another feed additive named E-BIOTICTM equine in USA contained *Lb. lactic*; *Lb. acidophilus*; *Lb. plantarum* and other LAB [113].

4. Lactobacilli in the pharmaceutical industry

The use of *Lactobacilli* has frequently been associated with the prevention of disease. It has been revealed from the literature that the oral administration of large dosages of lactobacilli is an effective and nontoxic in patients having Crohn's disease, AIDS and various types of enteritis, and in premature babies, children or the senior adults, all of those are possibly at risk of lactobacillemia [114–120]. In some cases, enteral administration of *Lactobacillus* species may be favorable [121] in rats and humans [122]. *Lactobacilli* may also have positive effects in the oral cavity, inhibiting cariogenic *Candida* and *streptococci* species [123]. Tallying to these trials, efficient medicines containing viable *Lactobacilli* with no side effects have developed (**Table 4**).

5. Beneficial role of *lactobacillus*

Lactobacilli can produce lactic acid in large amounts by a biotechnological process through fermentation. Among **lactobacilli**, lactic acid-producing bacteria play an important role in the production of fermented dairy products and the enzyme beta-galactosidase. Because these bacterial species are generally recognized as safe so enzyme purification does not require any extensive purification processes and also there is no damage to fermented food products [124].

A large number of Gram-negative and Gram-positive bacteria produce antimicrobial protein structures during their growth (either polypeptides or proteins) called bacteriocins [125]. Now a days Lactic acid bacteria have gained specific consideration among the Gram-positive bacterial strains because of the production of bacteriocins

Lactobacillus species	Trademark	Dose (Adult) – Comments
Lb. casei rhamnosus	Bacilors (347961.2)	Ingestion, 2 to 8 capsules/day, anti-diarrhea
Lb. acidophilus	Lacteols (305665.6)	Ingestion, 5 tablets/3 to 5 times/day, anti-diarrhea
Lb. rhamnosus	Antibiophiluss	Capsule ingestion, anti-diarrhea

Table 4.
List of drugs (capsule, tablet or gel) based on lactobacilli (http://www.vidalpro.net).

[126]. The bacteriocins can be functional in the food industry by employing as natural preservatives. The application of produced bacteriocins having antimicrobial compounds as a natural barrier against food spoilage causing pathogenic microbes has been confirmed to be effective [38]. The following approaches of bacteriocin application in food are usually adopted for the purpose of biopreservation: adding semi purified or pure bacteriocin as additive in food; food is inoculated with the bacteriocin-producing strain; and practice of previously fermented product with a strain having ability to produce bacteriocin as an ingredient in food processing [38]. A new application of bacteriocins is in bioactive packaging which is a procedure that can defend the food from exterior contaminants [126].

Lactobacilli having probiotic effects in humans are completely documented and are dose and strain dependent [127]. Numerous articles, reviews and book chapters have documented the potential lactobacilli to promote overall health: help to mitigate lactose intolerance [128], positively affect the flora of intestinal tract [129], prevent gastrointestinal infections [130, 131], stimulate the immune system [132], decrease in allergies or inflammation responses [133, 134], regulate the motility of gut [135], and promotion of sensation of well-being [136]. Particular health effects such as colon cancer prevention of [137–139] and decreases in blood lipid concentration and heart disease [140] antihypertensive effects have also been reported. **Figure 2** shows the use of probiotics in different food items.

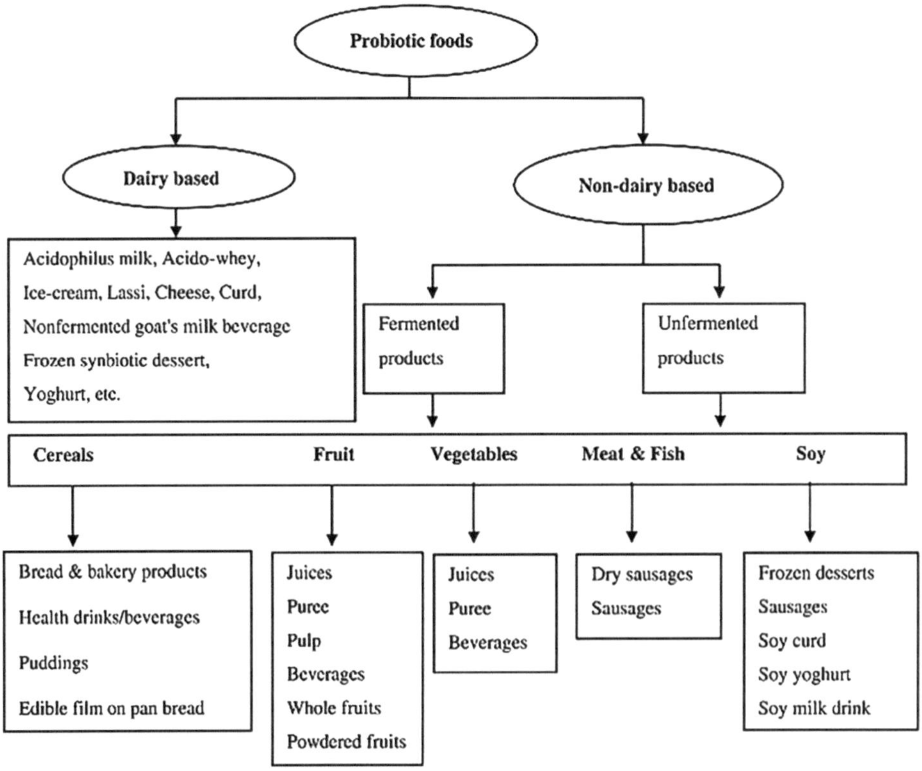

Figure 2.
Probiotic foods in which lactobacilli are applied as a probiotic strain.

DOI: http://dx.doi.org/10.5772/intechopen.106856

6. Contribution of lactobacilli to the United Nations development goals

Lactobacilli in the form of starter culture have prospective potential for application in foods of animal origin. The starter cultures are extensively used in in the meat industry for production of sausage, the dairy industry for yoghurt, cheese and other fermented dairy products and in the fishery industry for fermented fish products. The starter cultures are mainly used in food of animal origin for quick lactic acid production which causes a decrease in pH and thus preventing the growth of spoilage and foodborne pathogenic microbes which ultimately increasing the shelf-life of fermented food products. Also, release of other metabolic compounds (e.g., acetic acid, lactic acid, benzoic acid, propionic acid and hydrogen peroxide or bacteriocins) helps in improving the safety of foods [141]. Since starter cultures have developed as the predominant microbiota of fermented foods and allows the food processors to control the process of fermentation by eliminating the objectionable microflora and reducing the manufacturing and hygienic risks. Another important role of starter cultures is the chemical safety of fermented products/foods by decreasing aromatic polycyclic hydrocarbons and biogenic amine contents. Therefore, the futuristic application of lactobacilli and other microbes as starter culture crucially helps in reducing hunger by elimination of food spoilage which is the main cause of food insecurity and play a significant role in improving health and wellbeing by providing essential nutrients and reducing the population of undesirable microbiota and increasing the shelf life of foods. So the *lactobacillus* with other microbes can significantly effective for their progressive utilization in important areas of sustainable development goals agenda of 2030, the number one is huger and second is health and wellbeing. At the same time lactobacilli have huge industrial potential and it is the dire need of the day to explore more potential lactobacilli with multiple characteristics for food and pharmaceutical industry.

7. Conclusions

Lactobacilli, a potentially significant group of microorganisms that play a fundamental role in the fermentation industry for the production of numerous valuable products such as cheese, yoghurt, fermented milk, kefir, enzymes, medicines, etc. Because of the diverse applications *lactobacillus* genus open new doors for exciting research areas. The application of lactobacilli in feed, pharmaceuticals and probiotics ensures to reduce the rate of mortality and morbidity associated with gastroenteritis and viral infections. The extensive industrial application of Lactobacilli demanded further research to enhance the potential of the species of this group and to isolate efficient strains which can produce multiple products of commercial practice. *Lactobacilli* have the futuristic potential for utilization in a diversity of industrial applications and hence can be considered a significant candidate for poverty elimination, reducing hunger by producing a diverse range of food products, increasing the shelf life by preserving the food and providing valuable pharmaceutical products.

Conflict of interest

"The authors declared no conflict of interest."

Author details

Nagina Rafique*, Turfa Mamoona, Saiqa Bashir, Imtiaz Hussain and Imran Hayat
Faculty of Agriculture, Department of Food Science and Technology, University of Poonch, Rawalakot, Azad Kashmir, Pakistan

*Address all correspondence to: nagina_rafiq2008@yahoo.com; drnaginarafiq@upr.edu.pk

References

[1] Cannon JP, Lee TA, Bolanos JT, Danziger LH. Pathogenic relevance of lactobacillus: A retrospective review of over 200 cases. European Journal of Clinical Microbiology & Infectious Diseases. 2005;**24**:31-40. DOI: 10.1007/s10096-004-1253-y

[2] Madigan MT, John MM, Thomas DB. Prokaryotic diversity: The bacteria. In: Madigan MT, Martinko JM, editors. Brock: Biology of Microorganisms. Upper Saddle River, NJ: Prentice Hall; 2006. pp. 329-418

[3] Kononen E, Wade WG. Propionibacterium, lactobacillus, Actinomyces, and other non-sporeforming anaerobic gram-positive rods. In: Murray PR, editor. Manual of Clinical Microbiology. Washington, DC: ASM Press; 2007. pp. 872-888

[4] International Scientific Association for Probiotics and Prebiotics. New names for important probiotic lactobacillus species. In: ISAPP Science Blog. Sacramento: ISAPP 2020; 2020 [about 2 screens]. Available from: https://isappscience.org/new-names-for-important-probiotic-lactobacillus-species/

[5] Zheng J, Wittouck S, Salvetti E. A taxonomic note on the genus lactobacillus: Description of 23 novel genera, emended description of the genus lactobacillus Beijerinck 1901, and union of Lactobacillaceae and Leuconostocaceae. International Journal of Systematic and Evolutionary Microbiology. 2020;**70**:2782-2858. DOI: 10.1099/ijsem.0.004107

[6] Aguirre M, Collins MD. Lactic acid bacteria and human clinical infection. The Journal of Applied Bacteriology. 1993;**75**:95-107. DOI: 10.1111/j.1365-2672.1993.tb02753.x

[7] Sharpe ME, Hill LR, Lapage SP. Pathogenic lactobacilli. Journal of Medical Microbiology. 1973;**6**:281-286. DOI: 10.1099/00222615-6-3-281

[8] Husni RN, Gordon SM, Washington JA, Longworth DL. Lactobacillus bacteremia and endocarditis: Review of 45 cases. Clinical Infectious Diseases. 1997;**25**:1048-1055. DOI: 10.1086/516109

[9] Abgrall S, Joly V, Derkinderen P, Decré D, Carbon C, Yeni P. Lactobacillus casei infection in an AIDS patient. European Journal of Clinical Microbiologyv & Infectious Diseases. 1997;**16**:180-182. DOI: 10.1007/BF01709485

[10] Chomarat M, Espinouse D. Lactobacillus rhamnosus septicemia in patients with prolonged aplasia receiving ceftazidime-vancomycin. European Journal of Clinical Microbiology & Infectious Diseases. 1991;**10**:44. DOI: 10.1007/BF01967099

[11] Schlegel L, Lemerle S, Geslin P. Lactobacillus species as opportunistic pathogens in immunocompromised patients. European Journal of Clinical Microbiology & Infectious Diseases. 1998;**17**:887-888. DOI: 10.1007/s100960050216

[12] Harty DWS, Oakey HJ, Patrikakis M, Hume EB, Knox KW. Pathogenic potential of lactobacilli. International Journal of Food Microbiology. 1994;**24**:179-189. DOI: 10.1016/0168-1605(94)90117-1

[13] Inglin CR. High-throughput screening assays for antibacterial and

antifungal activities of lactobacillus species. Journal of Microbiological Methods. 2015;**114**:26-29. DOI: 10.1016/j.mimet.2015.04.011

[14] Holzapfel WH. Appropriate starter culture technologies for small-scale fermentation in developing countries. International Journal of Food Microbiology. 2002;**75**:197-212. DOI: 10.1016/s0168-1605(01)00707-3

[15] Konigs WN, Kok J, Kuipers OP, Poolman B. Lactic acid bacteria: The bug of the new millennium. Current Opinion in Microbiology. 2000;**3**:276-282. DOI: 10.1016/s1369-5274(00)00089-8

[16] Adams MR. Safety of industrial lactic acid bacteria. Journal of Biotechnology. 1999;**68**:171-178. DOI: 10.1016/s0168-1656(98)00198-9

[17] Marion Bernardeau M, Guguen M, Vernoux JP. Beneficial lactobacilli in food and feed: Long-term use, biodiversity and proposals for specific and realistic safety assessments. FEMS Microbiology Reviews. 2006;**30**:487-513. DOI: 10.1111/j.1574-6976.2006.00020.x

[18] Battcock M, Azam-Ali S, FAO. Fermented Fruits and Vegetables. A Global Perspective. Agricultural Support Systems D. Rome, Italy: FAO; 1998. Available from: http://www.fao.org

[19] Saloff-Coste CJ. *Lactobacillus casei*. Danone World Newsletter; 1995. pp. 2-8. Available from: http://www.danonevitapole.com/nutri_views/newsletter/eng/news_7

[20] Ross RP, Desmond C, Fitzerald GF, Stantch C. Overcoming the technological hurdles in the development of probiotic foods. Journal of Applied Microbiology. 2005;**98**:1410-1417. DOI: 10.1111/j.1365-2672.2005.02654.x

[21] Fox PF, Guinee TP, Cogan TM, McSweeney PLH. Pathogens in cheese and foodborne illnesses. In: Fundamental of Cheese Science. New York, NY, USA: Springer; 2017. pp. 681-713

[22] Hammami R, Fliss I, Corsetti A. Application of protective cultures and bacteriocins for food biopreservation. Frontiers in Microbiology. 2019;**1561**:594. DOI: 10.3389/fmicb.2019.01561

[23] Aryana KJ, Olson DWA. Yogurt and other cultured dairy products. Journal of Dairy Science. 2017;**100**:9987-10013

[24] Laranjo M, Elias M, Fraqueza MJ. The use of starter cultures in traditional meat products. Journal of Food Quality. 2017;**2017**:9546026. DOI: 10.1155/2017/9546026

[25] Romano P, Capece A. Wine microbiology. In: Speranza B, Bevilacqua A, Corbo MR, Sinigaglia M, editors. Starter Cultures in Food Production; Chapter 13. West Sussex, UK: John Wiley & Sons, Ltd.; 2017. pp. 255-282

[26] Vaughan A, O'Sullivan T, Van Sinderen D. Enhancing the microbiological stability of malt and beer—A review. Journal of the Institute of Brewing. 2005;**111**:355-371. DOI: 10.1002/j.2050-0416.2005.tb00221.x

[27] Mas A, Torija MJ, García-Parrilla MDC, Troncoso AM. Acetic acid bacteria and the production and quality of wine vinegar. Scientific World Journal. 2014;**2014**:394671. DOI: 10.1155/2014/394671

[28] Gao L, Yang H, Wang X, Huang Z, Ishii M, Igarashi Y, et al. Rice straw fermentation using lactic acid bacteria. Bioresource Technology. 2008;**99**:2742-2748. DOI: 10.1016/j.biortech.2007.07.001

[29] Ashaolu TJ, Reale AA. Holistic review on euro-Asian lactic acid bacteria fermented cereals and vegetables. Microorganisms. 2020;**8**:1176. DOI: 10.3390/microorganisms8081176

[30] Arroyo-López FN, Garrido-Fernández A, Jiménez-Díaz R. Starter cultures in vegetables with special emphasis on table olives. In: Speranza B, Bevilacqua A, Corbo RM, Sinigaglia M, editors. Starter Cultures in Food Production. West Sussex, UK: John Wiley and Sons; 2017. pp. 283-298

[31] Brandt MJ. Starter cultures for cereal based foods. Food Microbiology. 2014;**37**:41-43. DOI: 10.1016/j.fm.2013.06.007

[32] Szutowska J. Functional properties of lactic acid bacteria in fermented fruit and vegetable juices: A systematic literature review. European Food Research and Technology. 2020;**246**:357-372. DOI: 10.1007/s00217-019-03425-7

[33] Linares MB, Garrido MD, Martins C, Patarata L. Efficacies of garlic and *L. sakei* in wine-based marinades for controlling listeria monocytogenes and salmonella spp. in chouriço de vinho, a dry sausage made from wine marinated pork. Journal of Food Science. 2013;**78**:M719-M724. DOI: 10.1111/1750-3841.12094

[34] Sun Q, Sun F, Zheng D, Kong B, Liu Q. Complex starter culture combined with vacuum packaging reduces biogenic amine formation and delays the quality deterioration of dry sausage during storage. Food Control. 2019;**100**:58-66. DOI: 10.1016/j.foodcont.2019.01.008

[35] NCBI. Data Collected from Genbank. NCBI. Rockville Pike, Bethesda, MD, USA: National Center for Biotechnology Information, National Library of Medicine, National Institutes of Health; 2018. Available from: https:// www.ncbi.nlm.nih.gov/genome/browse/

[36] Jack RW, Tagg JR, Ray B. Bacteriocins of gram-positive bacteria. Microbiological Reviews. 1995;**59**:171-200

[37] Bernadeau M, Vernoux JP, Henri-Dubernet S, Gueguen M. The lactobacillus genus. International Journal of Food Microbiology. 2007;**126**:103-125

[38] Chen H, Hoover DG. Bacteriocins and their food applications. Comprehensive Reviews in Food Science and Food Safety. 2003;**2**:83-97. DOI: 10.1111/j.1541-4337.2003.tb00016.x

[39] Beresford T, Cogan T. Improving Cheddar cheese flavor. In: Proceedings of the 5th Cheese Symposium. Cork: Teagasc/University College Cork; 1997. pp. 53-61

[40] Picon A. Cheese microbial ecology and safety. In: Papademas P, Bintsis T, editors. Global Cheese Making Technology, Cheese Quality and Characteristics. Chichester: John Wiley & Sons Ltd.; 2018. pp. 71-99

[41] Upadhyay VK, McSweeney PLH, Magboul AAA. Proteolysis in cheese during ripening. In: Cheese: Chemistry, Physics and Microbiology, 4. London: Elsevier Academic Press; 2004. pp. 391-433

[42] Curtin AC, McSweeney PLH.: Catabolism of amino acids in cheese during ripening, In: Fox PF, McSweeney PLH, Cogan TM, et al., Editors, Cheese: Chemistry, Physics and Microbiology, London: Elsevier Academic Press; 2004. pp. 435-454

[43] Alamprese C, Foschino R, Rossi M, Pompei C, Savani L. Survival of lactobacillus johnsonii La1 and influence of its addition in retail-manufactured

ice-cream produced with different sugar and fat concentrations. Dairy Journal. 2002;**12**:201-208. DOI: 10.1016/S0958-6946(01)00159-5

[44] Bernet-Camard MF, Liévin V, Brassart D. The human lactobacillus acidophilus strain La1 secretes a non bacteriocin antibacterial substance active in vitro and in vivo. Applied and Environmental Microbiology. 1997;**63**:2747-2753. DOI: 10.1128/aem.63.7.2747-2753

[45] Sgouras DN, Panayotopoulou EG, Martinez-Gonzalez B, Petraki K, Michopoulos S, Mentis A. Lactobacillus johnsonii La1 attenuates helicobacter pylori associated gastritis and reduces levels of proinflammatory chemokines in C57BL/6 mice. Clinical and Diagnostic Laboratory Immunology. 2005;**12**:1378-1386. DOI: 10.1128/CDLI.12.12.1378-1386

[46] Pérez PF, Minnaard J, Rouvet M. Inhibition of Giardia intestinalis by extracellular factors from lactobacilli: An in vitro study. Applied and Environmental Microbiology. 2001;**67**:5037-5042. DOI: 10.1128/AEM.67.11.5037-5042.2001

[47] Link-Amster H, Rochat F, Saudan KY. Modulation of a specific humoral immune response and changes in intestinal flora mediated through fermented milk intake. FEMS Immunology And Medical Microbiology. 1994;**10**:55-64. DOI: 10.1111/j.1574-695X.1994.tb00011.x

[48] Schiffrin EJ, Rochat F, Link-Amster H. Immunomodulation of human blood cells following the ingestion of lactic acid bacteria. Journal of Dairy Science. 1995;**78**:491-497. DOI: 10.3168/jds.S0022-0302(95)76659-0

[49] Schleifer KH, Ludwig W. Phylogenetic relationships of lactic acid bacteria. In: The Genera of Lactic Acid Bacteria. London: Chapman & Hall; 1995. pp. 7-18

[50] Hassan AN, Frank JF. Starter cultures and their use. In: Marth EH, Steele JL, editors. Applied Dairy Microbiology, 2 Eds. New York: Marcel Dekker Inc.; 2001. pp. 151-206

[51] Hammes WP, Vogel RF. The genus lactobacillus. In: The Genera of Lactic Acid Bacteria. London: Blackie Academic and Professional; 1995. pp. 19-54. DOI: 10.1111/j.1574-6968.2006.00596.x

[52] Wouters JTM, Ayad EHE, Hugenholtz J, Smit G. Microbes from raw milk for fermented dairy products. International Dairy Journal. 2002;**12**:91-109. DOI: 10.1016/S0958-6946(01)00151-0

[53] Coppola R, Nanni M, Iorizzo M, Sorrentino A, Sorrentino E, Chiavari C, et al. Microbiological characteristics of Parmegiano Reggiano cheese during the cheese making and the first months of the ripening. Le Lait. 2000;**2000**:479-490. DOI: 10.1051/lait:2000139ff. ffhal-00895440

[54] Hebert EM, Raya RR, Tailliez P, de Giori GS. Characterization of natural isolates of lactobacillus strains to be used as starter cultures in dairy fermentation. International Journal of Food Microbiology. 2000;**59**:19-27. DOI: 10.1016/s0168-1605(00)00282-8

[55] Mannu L, Communian R, Scintu MF. Mesophilic lactobacilli in Fiore Sardo cheese: PCR-identification and evolution during cheese ripening. International Dairy Journal. 2000;**10**:383-389. DOI: 10.1016/S0958-6946(00)00074-1

[56] Depouilly A, Dufrene F, Beuvier E, Berthier F. Genotypic characterisation of the dynamics of the lactic acid bacterial population of Comte cheese.

Le Lait. 2004;**84**:155-167. DOI: 10.1051/lait:2003036ffhal-00895521

[57] Henri-Dubernet S. Etude par electrophore´ `se en gradient de temperature de la biodiversit ´ e et des dynamiques des ´ populations lactobacilles au cours de la transformation du lait cru en "Camembert de Normandie" [thesis]. Caen: Universite de´ Caen Basse-Normandie; 2004

[58] Psoni L, Tzanetakis N, Litopoulou-Tzanetaki E. Microbial characteristics of Batzos, a traditional Greek cheese from raw goat's milk. Food Microbiology. 2003;**20**:575-582. DOI: 10.1016/S0740-0020(02)00153-3

[59] Coppola R, Blaiotta G, Ercolini D, Moschetti G. Molecular evaluation of microbial diversity occurring in different types of mozzarella cheese. Journal of Applied Microbiology. 2001;**90**:414-420. DOI: 10.1046/j.1365-2672.2001.01262.x

[60] Baruzzi F, Morea M, Matarante A, Cocconcelli PS. Changes in the lactobacillus community during ricotta forte cheese natural fermentation. Journal of Applied Microbiology. 2000;**89**:807-814. DOI: 10.1046/j.1365-2672.2000.01183.x

[61] Codex alimentarius. Codex Standard for Fermented Milk. CODEX STAN; 243. Viale delle Terme di Caracalla, Rome, Italy: FAO USA; 2003. pp. 1-5

[62] Garrote GL, Delfederico L, Bibiloni R, Abraham AG, Perez PF, Semorile L, et al. Lactobacilli isolated from kefir grains: Evidence of the presence of S-layer proteins. The Journal of Dairy Research. 2004;**71**:222-230. DOI: 10.1017/S0022029904000160

[63] Alimentarius Codex. Milk and milk products (2nd ed.). Rome, Italy: WHO and FAO. Moreno, De, de LeBlanc A, Matar C, Farnworth E, & Perdigon G. Study of cytokines involved in the prevention of a murine experimental breast cancer by kefir. Cytokine. 2006, 2011;1-2:1-8

[64] http://www.valio.f

[65] Davidson RH, Duncan SE, Hackney CR, Eigel WN, Boling JW. Probiotic culture survival and implications in fermented frozen yogurt characteristics. Journal of Dairy Science. 2000;**83**:666-673. DOI: 10.3168/jds.S0022-0302(00)74927-7

[66] Hekmat S, McMahon DJ. Survival of lactobacillus acidophilus and Bifidobacterium bifidum in ice cream for use as a probiotic food. Journal of Dairy Science. 1992;**75**:1415-1422. DOI: 10.3168/jds.S0022-0302(92)77895-3

[67] Farahmand N, Ouoba LII, Raeisi SN, Sutherland J, Ghoddusi HB. Probiotic lactobacilli in fermented dairy products: Selective detection, enumeration and identification scheme. Microorganism. 2021;**2021**:2-16. DOI: 10.3390/microorganisms9081600

[68] Miller BM, Litsky W. Single Cell Protein in Industrial Microbiology. 1st ed. New York: McGrow-Hill Book Co; 1976. p. 408. DOI: ajft.2011.103.116

[69] Plengvidhya V, Breidt F, Fleming HP. Use of RAPD-PCR as a method to follow the progress of starter cultures in sauerkraut fermentation. International Journal of Food Microbiology. 2004;**93**:287-296. DOI: 10.1016/j.ijfoodmicro.2003.11.010

[70] Krockel L, Schillinger U, Franz CM, Bantleon A, Ludwig W. Lactobacillus versmoldensis sp. nov., isolated from raw fermented sausage. International Journal of Systematic and Evolutionary Microbiology. 2003;**53**:513-517. DOI: 10.1099/ijs.0.02387-0

[71] Caggia C, Randazzo CL, Di Salvo M, Romeo F, Giudici P. Occurrence of listeria monocytogenes in green table olives. Journal of Food Protection. 2004;**67**:2189-2194. DOI: 10.4315/0362-028X-67.10.2189

[72] Campaniello D, Bevilacqua A, D'Amato D, Corbo MR, Altieri C, Sinigaglia M. Microbial characterization of table olives processed according to Spanish and natural styles. Food Technology and Biotechnology. 2005;**43**:289-294

[73] Cocolin L, Rantsiou K, Iacumin L, Urso R, Cantoni C, Comi G. Study of the ecology of fresh sausages and characterization of populations of lactic acid bacteria by molecular methods. Applied and Environmental Microbiology. 2004;**70**:1883-1894. DOI: 10.1128/AEM.70.4.1883-1894.2004

[74] Huerta R, Jordano R, Medina LM, Lopez C. Population dynamics of the constitutive biota of French dry sausages in a pilot-scale ripening chamber. Journal of Food Protection. 2004;**67**:2306-2309. DOI: 10.4315/0362-028X-67.10.2306

[75] Fontana C, Sandro-Cocconcelli P, Vignolo G. Monitoring the bacterial population dynamics during fermentation of artisanal Argentinian sausages. International Journal of Food Microbiology. 2005;**103**:131-142. DOI: 10.1016/j.ijfoodmicro.2004.11.046

[76] Vogel RF, Knorr R, Muller MRA, Steudel UG, anzle MG, Ehrmann MA. Non-dairy lactic fermentations: The cereal world. Antonie Van Leeuwenhoek. 1999;**76**:403-411. DOI: 10.1023/A:1002089515177

[77] Rousselin-Rousvoal F. L'Institut Rosell met des probiotiques dans le chocolat. Process. 2005;**1215**:50-51

[78] Luckow T, Delahunty C. Which juice is 'healthier'? A consumer study of probiotic non-dairy juice drinks. Food Quality Preference. 2004;**15**:751-759. DOI: 10.1016/j.foodqual.2003.12.007

[79] Park JS, Shin E, Hong H. Characterization of lactobacillus fermentum PL9988 isolated from healthy elderly Korean in a longevity village. Journal of Microbiology and Biotechnology. 2015;**25**:1510-1518. DOI: 10.4014/jmb.1505.05015

[80] Bintsis T. Lactic acid bacteria: Their applications in foods. Journal of Bacteriology and Mycology. 2018;**6**:89-94. DOI: 10.15406/jbmoa.2018.06.00182

[81] Dragoljub C, Aleksandra R, Dragiša S, Nataša J, Ana V, Lato P, et al. Survival of wild strains of lactobacilli during Kombucha fermentation and their contribution to functional characteristics of beverage. Polish Journal of Food and Nutrition Sciences. 2019;**69**:407-415. DOI: 10.31883/pjfns/112276

[82] Virdis C, Sumby K, Bartowsky E, Jiranek V. Lactic acid bacteria in wine: Technological advances and evaluation of their functional role. Frontiers in Microbiology. 2021;**11**:1-16. DOI: 10.3389/fmicb.2020.612118

[83] Liu SQ. Practical implications of lactate and pyruvate metabolism by lactic acid bacteria in food and beverage fermentations. International Journal of Food Microbiology. 2003;**83**:115-131. DOI: 10.1016/s0168-1605(02)00366-5

[84] López-Seijas J, García-Fraga B, da Silva AF, Zas-García X, Lois LC, Gago-Martínez A. Evaluation of malolactic bacteria associated with wines from Albariño variety as potential starters: Screening for quality and safety. Food. 2020;**9**:9-99. DOI: 10.3390/foods9010099

[85] Spano G, Beneduce L, Perrotta C, Massa S. Cloning and characterization

of the hsp 18.55 gene, a new member of the small heat shock gene family isolated from wine lactobacillus plantarum. Research in Microbiology. 2005;**156**:219-224. DOI: 10.1016/j.resmic.2004.09.014

[86] Rodas AM, Ferrer S, Pardo I. Polyphasic study of wine lactobacillus strains: Taxonomic implications. International Journal of Systematic and Evolutionary Microbiology. 2005;**55**:197-207. DOI: 10.1099/ijs.0.63249-0

[87] Adesogan AT, Salawu MB. The effect of different additives on the fermentation quality, aerobic stability and in vitro digestibility of pea/ wheat bi-crop silages containing contrasting pea to wheat ratios. Grass and Forage Science. 2002;**57**:25-32. DOI: 10.1046/j.1365-2494.2002.00298.x

[88] Seale DR. Bacterial inoculants as silage additives. The Journal of Applied Bacteriology. 1986;**8**:9S-26S

[89] Kung L, Ranjit NK. The effect of lactobacillus buchneri and other additives on the fermentation and aerobic stability of barley silage. Journal of Dairy Science. 2001;**84**:1149-1155. DOI: 10.3168/jds.S0022-0302(01)74575-4

[90] Hristov AN, McAllister TA. Effects of inoculants on whole-crop barley silage fermentation and dry matter disappearance in situ. Journal of Animal Science. 2002;**80**:510-516. DOI: 10.2527/2002.802510x

[91] Holzer M, Mayrhuber E, Danner H, Braun R. The role of lactobacillus buchneri in forage preservation. Trends in Biotechnology. 2003;**21**:282-287. DOI: 10.1016/S0167-7799(03)00106-9

[92] Nishino N, Wada H, Yoshida M, Shiota H. Microbial counts, fermentation products, and aerobic stability of whole crop corn and a total mixed ration ensiled with and without inoculation of lactobacillus casei or lactobacillus buchneri. Journal of Dairy Science. 2004;**87**:2563-2570. DOI: 10.3168/jds.S0022-0302(04)73381-0

[93] Schwenninger SM, Von Ah U, Niederer B, Teuber M, Meile L. Detection of antifungal properties in lactobacillus paracasei ssp. paracasei SM20, SM29, and SM63 and molecular typing of the strains. Journal of Food Protection. 2005;**68**:111-119. DOI: 10.4315/0362-028x-68.1.111

[94] Filya I, Sucu E, Karabulut A. The effect of Propionibacterium acidipropionici, with or without lactobacillus plantarum, on the fermentation and aerobic stability of wheat, sorghum and maize silages. Journal of Applied Microbiology. 2004;**97**:818-826. DOI: 10.1111/j.1365-2672.2004.02367.x

[95] Adesogan AT, Salawu MB, Ross AB, Davies DR, Brooks AE. Effect of lactobacillus buchneri, lactobacillus fermentum, Leuconostoc mesenteroides inoculants, or a chemical additive on the fermentation, aerobic stability, and nutritive value of crimped wheat grains. Journal of Dairy Science. 2003;**86**:1789-1796. DOI: 10.3168/jds.S0022-0302(03)73764-3

[96] Krooneman J, Faber F, Alderkamp AC, Elferink SJ, Driehuis F, Cleenwerck I, et al. Lactobacillus diolivorans sp. nov., a 1,2-propanediol-degrading bacterium isolated from aerobically stable maize silage. International Journal of Systematic and Evolutionary Microbiology. 2002;**52**:639-646. DOI: 10.1099/00207713-52-2-639

[97] Brambilla G, De Filippis S. Trends in animal feed composition and the possible consequences on residue tests. Analytica Chimica Acta. 2005;**2005**:7-13. DOI: 10.1016/j.aca.2004.07.067

[98] Simon O, Jadamus A, Vahjen W. Probiotic feed additives–effectiveness and expected modes of action. Journal of Animal and Feed Sciences. 2001;**10**:51-67. DOI: 10.22358/jafs/70012/2001

[99] La Ragione RM, Narbad A, Gasson MJ, Woodward MJ. In vivo characterization of lactobacillus johnsonii FI9785 for use as a defined competitive exclusion agent against bacterial pathogens in poultry. Letters in Applied Microbiology. 2004;**38**:197-205. DOI: 10.1111/j.1472-765x.2004.01474.x

[100] Li Z, Weiwei W, Dan L, Yuming G. Effects of lactobacillus acidophilus on the growth performance and intestinal health of broilers challenged with Clostridium perfringens. Journal of Animal Science and Biotechnology. 2018;**9**:9-25. DOI: 10.1186/s40104-018-0243-3

[101] Huang MK, Choi YJ, Houde R, Lee JW, Lee B, Zhao X. Effects of lactobacilli and an acidophilic fungus on the production performance and immune responses in broiler chickens. Poultry Science. 2004;**83**:788-795. DOI: 10.1093/ps/83.5.788

[102] Yoruk MA, Gul M, Hayirli A, Macit M. The effects of supplementation of humate and probiotic on egg production and quality parameters during the late laying period in hens. Poultry Science. 2004;**83**:84-88. DOI: 10.1093/ps/83.1.84

[103] Younts-Dahl SM, Osborn GD, Galyean ML, Rivera JD, Loneragan GH, Brashears MM. Reduction of Escherichia coli O157 in finishing beef cattle by various doses of lactobacillus acidophilus in direct-fed microbials. Journal of Food Protection. 2005;**68**:6-10. DOI: 10.4315/0362-028x-68.1.6

[104] Timmerman HM, Mulder L, Everts H, van Espen DC, Van der Wal E, Klaassen G, et al. Health and growth of veal calves fed milk replacers with or without probiotics. Journal of Dairy Science. 2005;**88**:2154-2165. DOI: 10.3168/jds.S0022-0302(05)72891-5

[105] Jaquette RD, Dennis RJ, Coalson J, Ware DR, Manfredi ET, Read PL. Effect of feeding viable lactobacillus acidophilus (BT1386) on the performance of lactating dairy cows. Journal of Dairy Science. 1988;**71**:219

[106] Kung L, Taylor CC, Lynch MP, Neylon JM. The effect of treating alfalfa with lactobacillus buchneri 40788 on silage fermentation, aerobic stability, and nutritive value for lactating dairy cows. Journal of Dairy Science. 2003;**86**:336-343. DOI: 10.3168/jds.S0022-0302(03)73611-X

[107] FAO. Poultry Development Review. Rome: FAO; 2013

[108] Lema M, Williams L, Rao DR. Reduction of fecal shedding of enterohemorrhagic Escherichia coli O157:H7 in lambs by feeding microbial feed supplement. Small Ruminant Research. 2001;**39**:31-39. DOI: 10.1016/s0921-4488(00)00168-1

[109] (http://www.efsa.eu.int)

[110] (D2/CSL http://www.efsa.eu.int)

[111] (DSM 13241 http://www.efsa.eu.int)

[112] (http://www.agrinutrition.com)

[113] (http://wysong.net)

[114] Donohue D, Salminen S. Safety of probiotic bacteria. Asia Pacific Journal of Clinical Nutrition. 1996;**5, 5**:25-28

[115] Salminen S, von Wright A, Laine M, Vuopio-Varkila J, Korhonen T, Mattila-Sandholm T. Development of

selection criteria for probiotic strains to assess their potential in functional foods: A Nordic and European approach. Bioscience and Microflora. 1996;**1996**:61-67. DOI: 10.12938/bifidus1996.15.61

[116] Salminen S, Von Wright A, Morelli L, Marteau P, Brassart D, De Vos R, et al. Demonstration of safety of probiotics–a review. International Journal of Food Microbiology. 1998;**44**:93-106. DOI: 10.1016/s0168-1605(98)00128-7

[117] Salminen MK, Tynkkynen S, Rautelin H, Saxelin M, Vaara M, Ruutu P, et al. Lactobacillus bacteremia during a rapid increase in probiotic use of lactobacillus rhamnosus GG in Finland. Clinical Infectious Diseases. 2002;**35**:1155-1160. DOI: 10.1086/342912

[118] Wolf BW, Wheeler KB, Ataya DG, Garleb KA. Safety and tolerance of lactobacillus reuteri supplementation to a population infected with the human immunodeficiency virus. Food and Chemical Toxicology. 1998;**36**:1085-1094. DOI: 10.1016/s0278-6915(98)00090-8

[119] Cunningham-Rundles S, Ahrne S, Bengmark S, Johann-Liang R, Marshall F, Metakis L, et al. Probiotics and immune response. The American Journal of Gastroenterology. 2000;**95**:22-S25. DOI: 10.1016/s0002-9270(99)00813-8

[120] Kullen MJ, Bettler J. The delivery of probiotics and prebiotics to infants. Current Pharmaceutical Design. 2005;**11**:55-74. DOI: 10.2174/1381612053382359

[121] Thorlacius H, Nobaek S, Wang XD, Andersson R, Molin G, Bengmark S, et al. Lactobacilli attenuate bacteremia and endotoxemia associated with severe intraabdominal infection. Surgery. 2003;**134**:467-473. DOI: 10.1067/s0039-6060(03)00246-0

[122] McNaught CE, Woodcock NP, Anderson AD, MacFie J. A prospective randomised trial of probiotics in critically ill patients. Clinical Nutrition. 2005;**24**:211-219. DOI: 10.1016/j.clnu.2004.08.008

[123] Meurman JH. Probiotics: Do they have a role in oral medicine and dentistry? European Journal of Oral Sciences. 2005;**113**:188-196. DOI: 10.1111/j.1600-0722.2005.00191.x

[124] Zacharof MP, Lovitt RW, Ratanapongleka K. The importance of lactobacilli in contemporary food and pharmaceutical industry a review article. In: Conference Paper Proceedings of 2010 International Conference on Chemical Engineering and Applications (CCEA 2010). Swansea, UK: Multidisciplinary Nanotechnology Center, Swansea University; 2010. pp. 13-18

[125] Beasley SS, Saris PEJ. Nisin-producing *Lactococcus lactis* strains isolated from human Milk. Applied and Environmental Microbiology. 2004;**70**:5051-5053. DOI: 10.1128/AEM.70.8.5051-5053.2004

[126] Ross RP, Morgan S, Hill C. Preservation and fermentation: Past, present and future. International Journal of Food Microbiology. 2002;**79**:3-16. DOI: 10.1016/s0168-1605(02)00174-5

[127] Gill HS, Guarner F. Probiotics and human health: A clinical perspective. Postgraduate Medical Journal. 2004;**80**:516-526. DOI: 10.1136/pgmj.2003.008664

[128] Levri KM, Ketvertis K, Deramo M, Merenstein JH, D'Amico F. Do probiotics reduce adult lactose intolerance? A systematic review. The Journal of Family Practice. 2005;**54**:613-620

[129] Qin HL, Shen TY, Gao ZG, Fan XB, Hang XM, Jiang YQ, et al. Effect of

lactobacillus on the gut microflora and barrier function of the rats with abdominal infection. World Journal of Gastroenterology. 2005:2591-2596. DOI: 10.3748/wjg.v11.i17.2591

[130] Macfarlane GT, Cummings JH. Probiotics, infection and immunity. Current Opinion in Infectious Diseases. 2002;**15**:501-506. DOI: 10.1097/00001432-200210000-00008

[131] Reid G, Burton J. Use of lactobacillus to prevent infection by pathogenic bacteria. Microbes and Infection. 2002;**4**:319-324. DOI: 10.1016/s1286-4579(02)01544-7

[132] Cross ML. Microbes versus microbes: Immune signals generated by probiotic lactobacilli and their role in protection against microbial pathogens. FEMS Immunology and Medical Microbiology. 2002;**34**:245-253. DOI: 10.1111/j.1574-695X.2002.tb00632.x

[133] Bongaerts GP, Severijnen RS. Preventive and curative effects of probiotics in atopic patients. Medical Hypotheses. 2005;**64**:1089-1092. DOI: 10.1016/j.mehy.2004.10.018

[134] Viljanen M, Savilahti E, Haahtela T, Juntunen-Backman K, Korpela R, Poussa T, et al. Probiotics in the treatment of atopic eczema/dermatitis syndrome in infants: A double-blind placebo-controlled trial. Allergy. 2005;**60**:494-500. DOI: 10.1111/j.1398-9995.2004.00514.x

[135] Hamilton-Miller JM. Probiotics and prebiotics in the elderly. Postgraduate Medical Journal. 2004;**80**:447-451. DOI: 10.1136/pgmj.2003.015339

[136] Torriani S, Marzotto M. Probiotic as promoters of human wellbeing: Proven effects and quality assessment of selected cultures. Industria del Latte. 2003;**39**:29-46

[137] Roller M, Pietro Femia A, Caderni G, Rechkemmer G, Watzl B. Intestinal immunity of rats with colon cancer is modulated by oligofructose-enriched inulin combined with lactobacillus rhamnosus and Bifidobacterium lactis. The British Journal of Nutrition. 2004;**92**:931-938. DOI: 10.1079/bjn20041289

[138] Floch MH. Use of diet and probiotic therapy in the irritable bowel syndrome: Analysis of the literature. Journal of Clinical Gastroenterology. 2005;**39**:243-246. DOI: 10.1097/01.mcg.0000156104.67505.5b

[139] Usman HA. Hypocholesterolemic effect of lactobacillus gasseri SBT0270 in rats fed a cholesterol-enriched diet. The Journal of Dairy Research. 2001;**68**:617-624. DOI: 10.1017/s0022029901005179

[140] Liong MT, Shah NP. Acid and bile tolerance and cholesterol removal ability of lactobacilli strains. Journal of Dairy Science. 2005;**2005**:55-66. DOI: 10.3168/jds.S0022-0302(05)72662-X

[141] García-Díez J, Cristina S. Use of starter cultures in foods from animal origin to improve their safety. International Journal of Environmental Research and Public Health. 2021;**18**:2544. DOI: 10.3390/ijerph18052544

Chapter 7

Lactobacillus exopolysaccharide: An Untapped Biopolymer

Christopher Osita Eze, Dinebari Phillip Berebon, Thaddeus Harrison Gugu, Francis Ifeanyi Anazodo and James Ekemezie Okorie

Abstract

Lactobacillus spp. belongs to a class of bacteria known as lactic acid bacteria. This classification is because they are known to produce lactic acid as a major by-product of their metabolic activities. Most *Lactobacillus* spp. are generally regarded as safe (GRAS) bacteria. They also produce a bio-polymeric substance known as exopolysaccharide (EPS). The EPS are popular because of their wide potential medical and industrial applications. The wide application of the EPS in medicine and industry necessitates optimal production and recovery of these polymeric substances produced by *Lactobacillus* spp. In this book chapter, we aim to comprehensively discuss *Lactobacillus* EPS, its inherent properties, potential pharmaceutical and industrial applications. We also point to its contribution towards the achievement of the 3rd and 9th components of the United Nations Sustainable Development Goals which are to establish good health and wellbeing and to promote industrialization, innovation, and infrastructure respectively.

Keywords: *Lactobacillus*, exopolysaccharide, biopolymer, homopolysaccharide, heteropolysaccharide

1. Introduction

1.1 Overview of *Lactobacillus* as a genus

The genus *Lactobacillus* spp. are members of lactic acid bacteria (LAB) belonging to the family Lactobacillaceae, order Lactobacillales, class Bacilli, and phylum Firmicutes. For the past decades, the *Lactobacillus* genera have been revised and new genus names were assigned. From a taxonomic point of view, the genus *Lactobacillus* comprises 261 species (at March 2020) that are extremely diverse at phenotypic, ecological, and genotypic levels [1]. More than any genera of LAB, the genus *Lactobacillus* is generally regarded as safe (GRAS) and finds application in the food, dairy, cosmetic and pharmaceutical industries. The genus *Lactobacillus* is non-spore-forming, catalase-negative or pseudocatalase, oxidase negative, obligate saccharolytic rods or coccobacilli generally characterized by a low guanine and cytosine (GC) content

IntechOpen

of the genome although the upper limit of GC content reaches 59.2 mol % [2]. The availability of 16S rRNA gene sequence and genome data has ultimately unlocked the frontiers of knowledge into the evolutionary relationships of *Lactobacillus* species.

Based on the type of sugar fermentation pathway, lactobacilli can be categorized into three groups, (i) obligatelyhomofermentative, that produce only lactic acid from glucose as an end product of carbohydrate metabolism through the glycolysis or Embden-Meyerhof- Parnas (EMP) pathway; (ii) facultativelyheterofermentative, that produce a mixture of lactic and acetic acid, diacetyl, acetoin, and carbon dioxide as end products of carbohydrate metabolism via the glycolysis or the phosphoketolase pathway, and; (iii) obligatelyheterofermentative, that produce lactic and acetic acid, or ethanol, and CO_2 as end products of carbohydrate metabolism via the phosphoketolase (6-phosphogluconate) pathway [3].

The lactobacilli have varied resistance to different NaCl concentrations, antibiotics, and temperature or pH range mostly due to cellular fatty acids, isoprenoidquinones, and other characteristics of their cell wall composition [4].

Until recent times, most studies on lactobacilli were focused on their application in food fermentation and formed the greater percentage of probiotics in current use. However, the production of exopolysaccharides biopolymers has added a new dimension to the usefulness of lactobacilli.

This book chapter aims to critically look at *Lactobacillus* EPS, its inherent properties, potential pharmaceutical and industrial applications, and how to harness it in line with the 3rd and 9th components of the United Nations Sustainable Development Goals which include the establishment of good health and wellbeing and to promote industrialization, innovation, and infrastructure respectively.

2. Composition of *Lactobacillus* exopolysaccharide (EPS)

The bioproduction of exopolysaccharides is of universal occurrence among eukaryotic (plants, seaweeds) and prokaryotic (bacteria) organisms. Microbial exopolysaccharides have many uses in numerous fields including food industries, farming, textiles, cosmetics, bioremediation and therapeutics, and pharmacy because of their different composition, structure, physical and chemical properties [5, 6].

Based on the type of monosaccharide monomers, EPS are divided into two groups of homopolysaccharides and heteropolysaccharides. The *Lactobacillus* spp. like other LAB can synthesize homopolysaccharides or heteropolysaccharides. The synthesized homopolysaccharides are glucans or fructans, which are composed of only one type of monosaccharide (glucose or fructose, respectively), whereas the heteropolysaccharides contain different types of monosaccharides in different proportions [7].

The homopolysaccharides EPS of lactobacilli with a molecular weight of greater than 106 Dalton are either branched or unbranched and composed of either glucose or fructose and are categorized into α-D-glucans (Dextran, Mutan, Alternan, and Reuteran), β-D-glucans, fructans (Levan and Inulin) and polygalactans [6, 8, 9]. Similarly, heteropolysaccharides EPS of lactobacilli have a molecular weight ranging from 104 and 6.0 × 106 Dalton and are made up of common sugars (D-glucose, D-galactose), rare sugars (L-rhamnose, mannose, arabinose, xylose, fucose, N-acetylglucosamine, N-acetylgalactosamine or glucuronic acid) and non-carbohydrate components (acetate, phosphate, sulfate, pyruvate, propionate, glycerate, amino acid, L- glutamate, and succinate [6, 9, 10]. Some examples of heteropolysaccharides EPS of lactobacilli are gellan, xanthan, and kefiran [6].

3. Importance of *Lactobacillus* exopolysaccharides (EPS)

3.1 Health benefits of *Lactobacillus* exopolysaccharide (EPS)

Lactobacillus exopolysaccharides though are produced to help the bacteria withstand unfavorable environmental conditions such as desiccation, toxic materials, and osmotic stress [11], their health benefits are far-reaching. These health benefits include.

3.1.1 Lactobacillus *EPS as rotavirus therapeutic agent or oral vaccine adjuvants*

Several studies have shown the importance of *Lactobacillus* EPS in rotavirus-induced diarrhea in children. *Lactobacillus* has shown an ability to suppress the replication of rotavirus through the improvement of the intestinal barrier [12]. This attribute may be linked to the EPS produced by the bacteria. The study by Kim et al. [13], showed a potential rotavirus therapeutic/vaccine adjuvant effect from EPS produced by *Lactobacillus plantarum*. The EPS according to the murine model study caused a reduction in the duration of diarrhea, reduced lesions of the epithelium, reduced replication of the rotavirus in the intestine, and a reduction in the time of recovery of the suckling mice [13].

3.1.2 Antioxidant property

Basically, free radicals are atoms that can damage cells leading to sickness and aging. Antioxidants are substances that can reduce the reactivity of free radicals through the donation of electrons. A study by Adebayo-Tayo and Fashogbon 2020, revealed that exopolysaccharide from *Lactobacillus.*

delburecki subsp. bulgaricus had an antioxidant activity that is comparable to ascorbic acid [14]. Other studies by Tang et al. [15], Silva et al. [16], and Yang et al. [17] all show potential antioxidant effects of *Lactobacillus* exopolysaccharide [15–17]. The antioxidant potential of EPS may be due to the bioactive component in its moiety that is capable of donating hydrogen ions [18].

3.1.3 Anti-cancer activity

Cancer is an abnormal growth of cells with consequent destruction of other cells and organs leading to death. Cancerous conditions are usually serious medical conditions that require serious attention. Methods for the treatment of cancer which include the use of chemotherapeutic agents, radiation, and surgery are invasive. This is because chemotherapeutics and radiation treatments are non-selective as they destroy both normal and cancerous cells leading to serious adverse effects. Based on these explanations, emphasis is now on natural products with anti-cancer activity as they are most likely to come with minimal adverse effects when compared with other agents. The EPS of *Lactobacillus* has been studied for potential anticancer activity with a lot of promising results. EPS of *Lactobacillus gasseri* showed good anti-proliferative activity against cervical cancer cell lines [19]. EPS of *L. plantarum* caused an increased expression of pro-apoptotic genes in mouse intestinal epithelial cancer [20]. *Lactobacillus kefri* and other *Lactobacillus* strains have shown activity in colorectal cancer [21, 22]. The mechanisms of anti-cancer activity of these EPS are postulated to be through the improvement of immunity, prevention of tumorigenesis, and induction of cancer cell apoptosis [23].

3.1.4 Antimicrobial properties of EPS

Probiotic bacteria are known to produce antimicrobial substances such as bacteriocin, organic acids, etc. Research from several scientists has shown that *Lactobacillus* EPS possesses antimicrobial properties. *In vitro* study of the antimicrobial property of EPS from *Lactobacillus rhamnosus* isolated from human breast milk showed significant activity against *E.coli* and *Salmonella typhimurium* [24]. Studies by other scientists confirmed the potential antimicrobial properties of EPS from other *Lactobacillus* spp. [25, 26].

3.1.5 Antiviral activity

The improvement of intestinal health through the use of immunobiotics is quite trending at the moment. Immunobiotics are supplements that contain immunoglobulin combined with probiotics and prebiotics. Immunobiotics protection against viral infection is through enhancement of innate and adaptive immunity that leads to the reduction of the duration of the disease, number of episodes, and viral shedding [27]. Pattern recognition receptors through interaction with EPS allow communication between the immunobiotics and the host. EPS of *Lactobacillus delbrueckii* showed an improved antiviral activity [28]. *L. plantarum* antirotavirus activity has been mentioned earlier [13]. Anti-adenovirus activity from EPS of *Lactobacillus, Leuconostoc,* and *Pediococcus* spp. has been recorded [29]. Another Study using a Swine testicular cell line, showed an inhibitory effect on transmissible gastroenteritis coronavirus infection by EPS from *L. plantarum* [30].

3.1.6 Other medical importance

These include anti-inflammatory [31], anti-cholesterol [32], Immunostimulatory [33], anti-diabetic properties [34].

3.2 The importance of *Lactobacillus* EPS in the food and dairy industry

3.2.1 EPS as a bioflocculant

Flocculants are agents that can cause the aggregation of dispersed particles in a suspension [35]. This makes the suspended particles easy to remove thus presenting a good mouthfeel of the food product. These flocculants can be grouped into inorganic, organic, and bioflocculants [36]. The use of inorganic and organic flocculants is associated with biological toxicities [36, 37]. These biological toxicities are of serious concern and gave birth to more reliable and biologically friendly bioflocculants. The bioflocculants have the advantages of non-toxicity, biodegradability, and no residual pollution [38]. Studies on the use of *Lactobacillus* exopolysaccharides as a bioflocculant have been carried out with promising results [16, 39].

3.2.2 EPS as biothickner and gelling agent

Thickeners and gelling agents are usually polysaccharides or protein derivatives. They are usually added to food to increase viscosity and stability while maintaining other desired characteristics of the food product. They are used extensively in dairy and non-dairy products. The use of *Lactobacillus* EPS as biothickners/gelling agents is shown by the works of [1, 40–42]. *Lactobacillus* EPS are widely used in the dairy industry as thickeners and texturizers with a major function of stabilizing the milk and its constituents [43].

3.2.3 EPS as a meat preservative and quality enhancer

The use of lactic acid bacteria to preserve meat is an old practice that has extensively helped in the sustenance of the food industry. These bacteria carry out this preservation through the production of antibacterial substances such as acetic acid, lactic acid, and bacteriocin. They also compete with other food spoilage pathogens for available nutritional substances hence the survival of the fittest. Another wonderful attribute of Lactic acid bacteria is their ability to produce EPS. This EPS is known to protect the bacteria against environmental stress such as high salt concentration, low PH, and low water activity [44]. This protection is known to sustain the preservative ability of LAB. The EPS is also known to reduce fat in the meat. The health importance of reduced fat in meat cannot be over-emphasized. This has led to a growing demand by consumers for fat-reduced meat. A study by Hilbig et al. [45] reported that EPS matrices formed by LAB in certain German meat products were able to cause a fat reduction and improved the quality of the meat product [45].

S/N	Organism	Medical/industrial applications	Referrences
1.	*Lactobacillus plantarum*	Cholesterol lowering property	Dilna et al. [32]
2.	*Lactobacillusdelbrueckii* subsp. *Bulgaricus* and *Lactobacillus acidophilus*	Improve the texture of fermented food	Vinogradov et al. [49]; Charchoghlyan et al. [50]
3.	*Lactobacillus johnsonii* and *Lactobacillus casei*	Immunomodulatory property	Gorska et al. [51]; Xiu et al. [52]
4.	*Lactobacillus rhamnosus* and *Lactobacillus gasseri*	Antimicrobial property	RiazRajoka et al. [24]; Parveen et al. [25]
5.	*Lactobacillusdelbrueckii*and *L. plantarum*	Antiviral property	Kanmani et al. [28]; Type et al. [53]; Yang et al. [30]
6.	*L. gasseri,Lactobacillus brevis,* and *Lactobacillus kefri*	Anticancer	Sungur et al. [19]; RiazRajoka et al. [21]; RiazRajoka et al. [54]
7.	*Lactobacillus paraplantarum*	Anti-inflammatory property	Pecikoza et al. [55]
8.	*Lactobacillus delbrueckii*subspbulgaricus	Antioxidant property	Tang et al. [15]
9.	*L. casei*	Viscocity enhancer in food industry	Zheng et al. [56]
10	*L. plantarum*	Enhancer of thr rheological property of food	Silva et al. [16]

Table 1.
The medical and industrial application of Lactobacillus *EPS.*

3.2.4 Exopolysaccharide as an inhibitor of syneresis

Syneresis is the expulsion of a liquid from a gel. It has been shown to affect food and food products negatively. This phenomenon known as syneresis should be minimized in food products without affecting the inert property of the food product. *Lactobacillus* exopolysaccharide according to the study was able to prevent syneresis in starch [46]. Other studies by Lynch et al. [47] and Han et al. [48] gave vivid importance to the syneresis preventive ability of exopolysaccharides produced by lactic acid bacteria [47, 48]. The exopolysaccharide's ability to prevent syneresis is because of its inert high water-binding affinity. This will then make water be retained in the food or dairy product (**Table 1**).

4. Challenges facing the industrial applications of EPS

4.1 Poor yield of the EPS by *Lactobacillus* spp.

The major challenge facing the industrial and medical applications of *Lactobacillus* exopolysaccharides is the poor production of the polymer by the bacteria. It is therefore very important that critical and intensified scientific experiments geared towards solving this problem be prioritized. This is because of the obvious importance of this EPS in medicine and industry. We strongly believe that the ability of the scientific community to solve this problem of poor yield rest on intensified scientific research. However, several types of research have been conducted and some are ongoing towards finding ways of improving the yield of this very important polymer [57–60].

4.2 Difficulty in isolation and characterization

The methods of extraction and characterization pose serious challenges to the industrial applications of EPS. There are several methods for the extraction of EPS from LAB. However, the physico-chemical properties of this polymeric substance can be affected by the method adopted [61]. The extraction and isolation of EPS usually involve the use of organic solvents, enzymes, sugars, filtration, chromatographic procedures, dialysis, etc. [62]. In all these, the most important thing is to target a high yield of pure EPS utilizing cheaper processes. To achieve this some important aspects of the process are taken into consideration namely the removal of proteins to avoid co-precipitation with EPS, and the prevention of reaction of the EPS with the solvents and components of the medium [61]. Since it is common knowledge that the major challenge of full industrial utilization of EPS is poor yield, there is, therefore, a need to adopt the best strategy for maximal extraction and characterization of the small quantity released by the bacteria. However, this process according to Badel et al. [63] is very laborious and indeed challenging [63].

5. Ways of mitigating the problems associated with EPS production by *Lactobacillus* spp.

A critical look at available scientific literature has shown that the importance of *Lactobacillus* EPS in medicine and industries cannot be overemphasized. It is therefore important that serious research and strategies targeted at improving the yield and manipulating its structures to achieve desired properties be the focus of scientists.

5.1 Screening for high EPS yielding strains

Badel et al. [63] in their study suggested that about 165 spp. of *Lactobacillus* are capable of EPS production [63]. Having identified these spp., there is then a need for extensive screening of these spp. to identify strains with potentially high EPS yielding capacity. It is a known fact that EPS are produced by these organisms to help them withstand environmental stress. Based on this fact it is, therefore, possible to experimentally create some of these stress conditions in the laboratories to enhance EPS production by the potentially high yielding strains. The stress may be in the form of deprivation or excessive introduction of certain nutritional substances. Organisms are known to respond differently to these challenges. A typical example is shown with *Lactobacillus lactis* subsp. cremoris and *Lactobacillus delbreckii* subsp. bulgaricus. While nitrogen deprivation improved the yield of EPS by *L. lactis* subsp. *cremoris* [64], nitrogen enrichment caused an improved EPS production by *Lactobacillus delbreckii* subsp. *bulgaricus* [65]. Similarly, sugar stress has also been shown to affect EPS production by *Lactobacillus* spp. It has been established that an increase in the concentration of sucrose in various media for the growth of *Lactobacillus* spp. has resulted in a significant increase in EPS production by the organisms [66–68]. Other stress conditions that enhance EPS production include osmotic stress [66] and temperature stress [69]. The stress-induced by the presence of other organisms has also led to enhanced EPS production. Several studies have shown that co-cultivation of some strains of *Lactobacillus* spp. with *Saccharomyces cerevisiae* led to improved production of EPS [70–72].

5.2 Development of genetically engineered strain

The advent of biotechnology and molecular biology has led to finding solutions to certain biological and physiological problems. Genetic engineering being the modification of organisms or population of organisms through manipulation or recombination of DNA or other nucleic acid molecules has led to the discovery of important substances such as human insulin and important vaccines. Due to the increasing importance of EPS in medicine and industries, a good understanding of genes encoding EPS production and their biosynthetic pathways has become very necessary. These studies have shown that there are four pathways involved in EPS production namely wzy-dependent pathway, the ATP-binding ABC transporter pathway, the synthase-dependent pathway, and extracellular synthesis by sucrose-dependent pathway [73]. *Lactobacillus* spp. are known to produce EPS via the wzy-dependent pathway [74]. Genetic engineering technology can enhance EPS production through manipulation of carbon metabolism and regulation of the biosynthetic pathways for its production [75]. Genetic manipulations leading to overexpression of genes and gene knockouts have caused an increase and structural changes in the produced EPS with desired characteristics.

5.3 Exploration of efficient extraction method

The production of a high yield of EPS requires optimization of the media for the growth of the bacteria. This involves the addition of substances to the established media for the growth of the bacteria. These additional substances aimed at enhancing EPS production may pose serious problems during the recovery of the final products. Therefore, successful production can be said to rely substantially on the use of a medium that not only allow high yield but also one whose components do not interfere with EPS component, recovery, and quantification [76]. In choosing a media it is advisable to choose one whose components are defined with minimal interfering compounds. Based on the complexities (no of additional substances for enrichment) of different media the more laborious the extraction/purification of the EPS. In complex media, a lot of pre-treatments and treatments are usually carried out. It is important to state that the use of complex media for a supposed gain in yield should be weighed against the laborious processes of the final recovery. This is because the possible gain in yield could be lost in the high cost and laborious processes of recovery. At times simple media may give substantial yield without the rigorous processes of recovery involved in the use of complex media. In all, the optimal recovery of the produced EPS should be the desire of the whole process.

6. *Lactobacillus* EPS and the sustainable development goal 2030 agenda of the United Nations

The 2030 agenda of the United Nations sustainable development goals was birthed in 2015 as a fallout of an agreement reached by 195 countries. It is subdivided into 17 components namely (1) Elimination of Poverty (2) Erasing Hunger (3) Establishing Good Health and Well-Being (4) Provision of Quality Education (5) Enforcement of Gender Equality (6) Improving Clean Water and Sanitation (7) Growing Affordable and Clean Energy (8) Creation of Decent Work and Economic Growth (9) Increasing Industry, Innovation, and Infrastructure (10) Reduction of Inequality (11) Mobilizing Sustainable Cities and Communities (12) Influencing Responsible Consumption and Production (13) Organization of Climate Action (14) Developing Life Below Water (15) Advancement of Life On Land (16) Guarantying Peace, Justice, and Strong Institutions (17) Building Partnerships for the Goal. The 3rd and 9th component is of special interest to us as it concerns our biopolymer. The use of natural products in therapy is associated with a more patient-friendly outcome. They usually do not produce as many side effects as synthetic products [77]. The commercial availability of this biopolymer means we are to witness new pharmaceutical formulations of important agents containing the EPS. This development will establish good health and wellbeing through the use of safe and efficacious pharmaceutical products. The pharmaceutical industry will also experience an increase in new therapeutic products for various indications containing the EPS. The dairy and meat industries are not left out as the use of EPS will be made to be more popular.

7. Conclusion and future perspectives

The EPS produced by *Lactobacillus* spp. are important biopolymers whose potential medical, pharmaceutical and industrial applications are still underutilized. The major problem of this underutilization rest mainly on the poor yield of the biopolymer. The selection of potentially high-yielding strains, exploitation of biotechnological principles through genetic engineering, and exploration of efficient extraction processes through extensive research will surely mitigate this problem and help give the biopolymer the required boost it deserves. Scientists in the field of probiotics should make concerted efforts towards the commercialization of EPS. Here we mean making characterized and lyophilized EPS powder commercially available for medical, pharmaceutical and industrial applications, especially in the food industry. The commercial availability of a lyophilized EPS powder will be a welcome development, especially for scientists in the pharmaceutical and food industry who really understand the importance of this biopolymer.

Since EPS is a biopolymer, its use in therapeutic formulations will be more patient-friendly devoid of side effects associated with synthetic or semi-synthetic chemicals. Furthermore, the commercial availability of fully characterized biopolymeric substances of *Lactobacillus* origin will create a huge innovation and industrial revolution in the pharmaceutical space this is because of new formulations of anticancer, antiviral, antidiabetic, anti-inflammatory, immunostimulants, cholesterol-lowering agents, etc. will be available for therapeutic applications. The same will be applicable in the dairy and meat industries as they will witness a reduction/complete elimination of the use of synthetic polymers. The full industrial and pharmaceutical application of EPS will contribute positively towards the establishment of improved health and wellbeing, as well as promote industrialization, drive innovation and increase infrastructure as envisaged in the United Nations Sustainable Development Goals.

Conflict of interest

The authors declare that there is no conflict of interest.

Author details

Christopher Osita Eze[1*], Dinebari Phillip Berebon[1], Thaddeus Harrison Gugu[1], Francis Ifeanyi Anazodo[1] and James Ekemezie Okorie[2]

1 Department of Pharmaceutical Microbiology and Biotechnology, University of Nigeria Nsukka, Enugu, Nigeria

2 Department of Pharmaceutical Technology and Industrial Pharmacy, University of Nigeria Nsukka, Enugu, Nigeria

*Address all correspondence to: osy.eze@unn.edu.ng

References

[1] Zheng J, Wittouck S, Salvetti E, CMAPM F, HMB H, Mattarelli P, et al. A taxonomic note on the genus *Lactobacillus*: Description of 23 novel genera, emended description of the genus *Lactobacillus Beijerinck* 1901, and union of Lactobacillaceae and Leuconostocaceae. International Journal of Systematic and Evolutionary Microbiology. 2020;**70**:2782-2858. DOI: 10.1099/ijsem.0.004107

[2] Cai Y, Pang H, Kitahara M, Ohkuma M. *Lactobacillus nasuensis* sp. nov., a lactic acid bacterium isolated from silage, and emended description of the genus *Lactobacillus*. International Journal of Systematic and Evolutionary Microbiology. 2012;**62**:1140-1144. DOI: 10.1099/ijs.0.031781-0

[3] Rossi F, Amadoro C, Colavita G. Members of the *Lactobacillus* genus complex (LGC) as opportunistic pathogens: A review. Microorganisms. 2019;**7**(5):126. DOI: 10.3390/microorganisms7050126

[4] Salvetti E, Torriani S, Felis GE. The genus *Lactobacillus*: A taxonomic update. Probiotics and Antimicrobial Proteins. 2012;**162**:853-865. DOI: 10.1007/s12602-012-9117-8

[5] Prado MR, Blandón LM, Vandenberghe LPS, Rodrigues C, Castro GR. Thomaz-Soccol V and Soccol CR Milk kefir: Composition, microbial cultures, biological activities, and related products. Frontiers in Microbiology. 2015;**6**:1177. DOI: 10.3389/fmicb.2015.01177

[6] Selim MS, Amer SK, Mohamed SS, Mounier MM, Rifaat HM. Production and characterisation of exopolysaccharide from *Streptomyces carpaticus* isolated from marine sediments in Egypt and its effect on breast and colon cell lines. Journal of Genetic Engineering and Biotechnology. 2017;**162**:853-865. DOI: 10.1016/j.jgeb.2017.10.014

[7] Angelin J, Kavitha M. Exopolysaccharides from probiotic bacteria and their health potential. International Journal of Biological Macromolecules. 2020;**162**:853-865

[8] Hussain A, Zia KM, Tabasum S, Noreen A, Ali M, Iqbal R. M. Zuber M. blends and composites of exopolysaccharides; properties and applications: A review, int. International Journal of Biological Macromolecules. 2017;**94**:10-27. DOI: 10.1016/j.ijbiomac.2016.09.104

[9] Torino MI, Valdez GF, Mozzi F. Biopolymers from lactic acid bacteria. Novel Applications in Foods and Beverages. 2015;**6**:1-16. DOI: 10.3389/fmicb.2015.00834

[10] Basiri S. Applications of microbial exopolysaccharides in the food Industry. Avicenna Journal of Medical Biochemistry. 2021;**9**(1):559. DOI: 10.34172/ajmb.2021.xx

[11] Wang CL, Huang TH, Liang TW, Fang CY, Wang SL. Production and characterization of exopolysaccharides and antioxidant from *Paenibacillus*sp. TKU023. New Biotechnology. 2011;**28**:559-565

[12] Mao X, Gu C, Hu H, Tang J, Chen DB, Yu B, et al. Dietary *Lactobacillus rhamnosus*GG supplementation improves the mucosal barrier function in the intestine of

weaned piglets challenged by porcine rotavirus. PLoS One. 2016;**11**:e0146312

[13] Kim K, Lee G, Thanh HD, Kim JH, Konkit M, Yoon S, et al. Exopolysaccharide from *Lactobacillu splantarum* LRCC5310 offers protection against rotavirus induced diarrhea and regulates inflammatory response. Journal of Dairy Science. 2018;**101**:5702-5712

[14] Adebayo-Tayo B, Fashogbon R. In vitro antioxidant, antibacterial, in vivo immunomodulatory, antitumor and haematological potential of exopolysaccharide produced by wild type and mutant *Lactobacillus delburecki*subspbulgaricus. Heliyon. 2020;**6**:e03268

[15] Tang W, Dong M, Wang W, Han S, Rui X, Chen X, et al. Structural characterization and antioxidant property of released exopolysaccharides from *Lactobacillus delbrueckii* ssp. bulgaricus SRFM 1. Carbohydrate Polymers. 2017;**173**:654-664

[16] Silva LA, Lopes Neto JHP, Cardarelli HR. Exopolysaccharides produced by *Lactobacillus plantarum*: technological properties, biological activity, and potential application in the food industry. Annals of Microbiology. 2019;**69**:321-328

[17] Yang SJ, Kim KT, Kim TY, Paik HD. Probiotic properties and antioxidant activities of *Pediococcus pentosaceus* Sc 28 and *Levilactobacillus brevis* KU15151 in fermented black Gamju. Food. 2020;**9**:1154

[18] Adebayo-Tayo B, Ishola R, Oyewunmi T. Characterization, antioxidant and immunomodulatory potential on exopolysaccharide produced by wild type and mutant *Weissellaconfusa* strains. Biotechnology Reports. 2018;**20**:e00271

[19] Sungur T, Aslim B, Karaaslan C, Aktas B. Impact of exopolysaccharides (EPSs) of *Lactobacillus gasseri* strains isolated from human vagina on cervical tumor cells (HeLa). Anaerobe. 2017;**47**:137-144

[20] Zhou X, Hong T, Yu Q, Nie S, Gong D, Xiong T. Exopolysaccharides from *Lactobacillus plantarum* NCU116 induce c-Jun dependent Fas / Fasl-mediated apoptosisvia TLR2 in mouse intestinal epithelial cancer cells. Scientific Reports. 2017;**45**:1-13

[21] RiazRajoka MS, Mehwish HM, Fang H, Padhiar AA, Zeng X, Khurshid M, et al. Characterization and anti-tumor activity of exopolysaccharide produced by *Lactobacilluskefiri* isolated from Chinese kefir grains. Journal of Functional Foods. 2019;**63**:103588

[22] Di W, Zhang L, Yi H, Han X, Zhang Y, Xin L. Exopolysaccharides produced by *Lactobacillus* strains suppress HT-29 cell growth via induction of G0/G1 cell cycle arrest and apoptosis. Oncology Letters. 2018;**16**:3577-3586

[23] Angelin J, Kavitha M. Exopolysaccharides from probiotic bacteria and their health potential. International Journal of Biological Macromolecules. 2020;**162**:853-865

[24] RiazRajoka MS, Jin M, Haobin Z, Li Q, Shao D, Jiang C, et al. Functional characterization and biotechnological potential of exopolysaccharide produced by *Lactobacillus rhamnosus* strains isolated from human breast milk, LWT - food sci. Technology. 2018;**89**:638-647

[25] Parveen R, Anandharaj M, David A. Internationaljournalofbiological MacromoleculesCharacterization of a novel exopolysaccharide produced by *Lactobacillus gasseri*FR4 anddemonstration of its in vitro

biological properties. International Journal of Biological Macromolecules. 2018;**109**:772-783

[26] Trabelsi I, Ktari N, Ben Slima S, Triki M, Bardaa S, Mnif H, et al. Evaluation of dermal wound healing activity and in vitro antibacterial and antioxidant activities of anew exopolysaccharide produced by *Lactobacillus* sp.Ca6. International Journal of Biological Macromolecules. 2017;**103**:94-201

[27] Angelin J, Kavitha M. Exopolysaccharides from probiotic bacteria and their health potential. International Journal of Biological Macromolecules. 2020;**162**:853-865

[28] KanmaniP,AlbarracinL,KobayashibH, Iida H, Komatsu R, Kober AKMH, et al. Exopolysaccharides from *Lactobacillus delbrueckii* OLL1073R-1 modulate innate antiviral immune response in porcine intestinal epithelial cells. Molecular Immunology. 2017:253-265

[29] Biliavska L, Pankivska Y, Povnitsa O, Zagorodnya S. Antiviral activity of exopolysaccharides produced by lactic acid bacteria of the genera *Pediococcus*, *Leuconostoc* and *Lactobacillus* against human adenovirus Type 5. Medicina (Kaunas, Lithuania). 2019;**55**(9):519

[30] Yang Y, Song HL, Wang L, Dong W, Yang Z, Yuan P, et al. Antiviral effects of a probiotic metabolic products against transmissible gastroenteritis coronavirus. Journal of Probiotics & Health. 2017;**5**(3):184

[31] Azad MAK, Sarkar M, Wan D. Immunomodulatory effects of probiotics on cytokine profiles. BioMed Research International. 2018:1-10. DOI: 10.1155/2018/8063647

[32] Dilna SV, Surya H, Aswathy RG, Varsha KK, Sakthikumar DN, et al. Characterization of an exopolysaccharide with potential health benefit properties from a probiotic *Lactobacillus plantarum*RJF4. LWT. Journal of Food Science and Technology. 2015;**64**:1179-1186

[33] Notararigo S, Las Casas-Engel MD, De Palencia PF, Corbi AL, Lopez P. 2014. Immunomodulation of human macrophages and myeloid cells by 2-substituted (1-3)-β-d-glucan from *P.Parvulus* 2-6. Carbohydr. Polymer. 2014;**112**:109-113

[34] Huang Z, Lin F, Zhu X, Zhang C, Jiang M, Lu Z. An exopolysaccharide from *Lactobacillus plantarum* H31 in pickled cabbage inhibits pancreas α-amylase and regulating metabolic markers in HepG2 cells by AMPK/P13K/Akt pathway. International Journal of Biological Macromolecules. 2020;**143**:775-784

[35] Kaarmukhilnilavan RS, Selvam A, Wong JW. MurugesanK Ca^{2+} dependent flocculation efficiency of avian egg protein revealed unique surface specific interaction with kaolin particles: A new perception in bioflocculant research. Colloids and Surfaces A: Physicochemical and Engineering Aspects. 2020;**603**:125177

[36] Salehizadeh H. Yan N (2014) recent advances in extracellular biopolymer flocculants. Biotechnology Advances. 2014;**32**(8):1506-1522

[37] Campbell A. The potential role of aluminium in Alzheimer's disease. Nephrol Dial Transplant. 2002;**17**(suppl_2):17-20

[38] Chaisorn W, Prasertsan P, Sompong O. Methacanon P (2016) production and characterization of biopolymer as bioflocculant from thermotolerant *Bacillus subtilis* WD161

in palm oil mill effluent. International Journal of Hydrogen Energy. 2016;**41**(46):21657-21664

[39] Wang Y, Ahmed Z, Feng W, Li C, Song S. Physicochemical properties of exopolysaccharide produced by *Lactobacillus kefiranofaciens* ZW3 isolated from Tibet kefir. International Journal of Biological Macromolecules. 2008;**43**(3):283-288

[40] Moradi Z, Kalanpour N. Kefiran, a branched polysaccharide: Preparation, properties and applications: A review. Carbohydrate Polymers. 2019;**223**:115100

[41] Rani RP, Anandharaj M, Ravindran AD. Characterization of a novel exopolysaccharide produced by *Lactobacillus gasseri* FR4 and demonstration of its *in vitro* biological properties. International Journal of Biological Macromolecules. 2018;**109**:772-783

[42] Gentès MC, Turgeon SL, St-Gelais D. Impact of starch and exopolysaccharide-producing lactic acid bacteria on the properties of set and stirred yoghurts. International Dairy Journal. 2016;**55**:79-86

[43] Korcz E, Varga L. Exopolysaccharides from lactic acid bacteria: Techno-functional application in the food industry. Trends in Food Science & Technology. 2021;**110**:375-384

[44] Rao MS, Pintado J, Stevens WF, Guyot JP. Kinetic growth parameters of different amylolytic and non-amylolytic *Lactobacillus* strains under various salt and pH conditions. Bioresource Technology. 2004;**94**:331-337

[45] Hilbig J, Gisder J, Prechtl RM, Herrmann K, Weiss J, Loeffler M. Influence of exopolysaccharide-producing lactic acid bacteria on the spreadability of fat-reduced raw fermented sausages (Teewurst). Food Hydrocolloids. 2019;**93**:422-431

[46] Ismail B, Nampoothiri KM. Exopolysaccharide production and prevention of syneresis in starch using encapsulated probiotic *Lactobacillus* plantarum. Food Technology and Biotechnology. 2010;**48**(4): 484-489

[47] Lynch KM, Zannini E, Coffey A, Arendt EK. Lactic acid bacteria exopolysaccharides in foods and beverages: Isolation, properties, characterization, and health benefits. Annual Review of Food Science and Technology. 2018;**9**:155-176

[48] Han X, Yang Z, Jing X, Yu P, Zhang Y, Yi H, Zhang L. Improvement of the texture of yogurt by use of exopolysaccharide producing lactic acid bacteria, BioMed Research International, 2016;**10**:7945675

[49] Vinogradov E, Sadovskaya I, Cornelissen A, van Sinderen D. Structural investigation of cell wall polysaccharides of *Lactobacillus delbrueckii* subsp bulgaricus 17. Carbohydrate Research. 2015;**413**:93-99

[50] Charchoghlyan H, Bae JE, Kwon H, Kim M. Rheological properties and volatile composition of fermented milk prepared by exopolysaccharide producing *Lactobacillus acidophilus* n.v.Er2 317/402 strain Narine. Biotechnology and Bioprocess Engineering. 2017;**22**:327-338

[51] Gorska S, Sandstrom C, Wojas-Turek J, Rossowska J, Pajtasz-Piasecka E, Brzozowska E, et al. Structural and immunomodulatory differences among lactobacilli exopolysaccharides isolated from intestines of mice with experimentally

induced inflammatory bowel disease. Scientific Reports. 2016;**6**:37613

[52] Xiu L, Zhang H, Hu Z, Liang Y, Guo S, Yang M, et al. Immunomodulatory activity of exopolysaccharides from probiotic *Lactobacilluscasei*. WXD 030 strain as a novel adjuvant *in vitro* and *in vivo*. *Food Agric*. Immunomodulator. 2018;**29**:1086-1105

[53] Type A, Biliavska L, Pankivska Y, Povnitsa O, Zagorodnya S. Antiviral activity of exopolysaccharides produced by lactic acid bacteria of the genera *Pediococcus, Leuconostoc,* and *Lactobacillus* against human adenovirus type 5. Medicina (Kaumas). 2019;**55**(9):519

[54] RiazRajoka MS, Mehwish HM, Zhang H, Ashraf M, Fang H, Zeng X, et al. Antibacterial and antioxidant activity of exopolysaccharide mediated silver nanoparticle synthesized by *Lactobacillus brevis* isolated from Chinese koumiss. Colloids Surface B Biointerfaces. 2020;**186**:110734

[55] Dinic M, Pecikoza U, Djokic J, Stepanovic-Petrovic R, Milenkovic M. Stevanovic M. et al. Exopolysaccharide produced by probiotic strain *Lactobacillus paraplantarum* BGCG11 reduces inflammatory hyperalgesia in rats. Frontiers in Pharmacology. 2018;**9**:1

[56] Zheng X, Yu Y, Xiao G, Xu Y, Wu J, Tang D, et al. Comparing product stability of probiotic beverages using litchi juice treated by high hydrostatic pressure and heat as substrates. Innovative Food Science and Emerging Technologies. 2014;**23**:61-67

[57] Oleksy-Sobczak M, Klewicka E. Optimization of media composition to maximize the yield of exopolysaccharides production by *Lactobacillus rhamnosus* strains. Probiotics and Antimicrobial Proteins. 2020;**12**:774-783

[58] Zhang L, Zhao B, Liu CJ, Yang E. Optimization of biosynthesis conditions for the production of exopolysaccharides by *Lactobacillus plantarum* SP8 and the exopolysaccharides antioxidant activity test. Indian Journal of Microbiology. 2020;**60**(3):334-345

[59] Benhadria MK, Tir TMA, Meddah B. Optimization of production of microbial exopolysaccharide (EPS) with essential oils from two medicinal plants. Journal of Applied Biosciences. 2017;**111**:10925-10933

[60] Liu Q, Huang X, Yang D, Si T, Pan S, Yang F. Yield improvement of exopolysaccharides by screening of the *Lactobacillus acidophilus* ATCC and optimization of the fermentation and extraction conditions. EXCLI Journal. 2016;**15**:119-133. DOI: 10.17179/excli2015-356

[61] Santos FL, Amorim GM. Biotechnological challenges and perspectives of using exopolysaccharides. Journal of Analytical & Pharmaceutical Research. 2018;7(3):264-266

[62] Rimada PS, Abraham AG. Comparative study of different methodologies to determine the exopolysaccharide produced by kefir grains in milk and whey. Le Lait. 2003;**83**:79-87

[63] Badel S, Bernardi T, Michaud P. New perspectives for *Lactobacillus* exoploysaccharide. Biotechnology Advances. 2011;**29**(1):54-66

[64] Marshall VM, Cowie EN, Moreton RS. Analysis and production of two exopolysaccharides from *Lactococcuslactis* subsp. *cremoris* LC330. The Journal of Dairy Research. 2009;**62**:621-628

[65] García-Garibay M, Marshall V. Polymer production by *Lactobacillus delbrueckii* ssp. Journal of Applied Microbiology. 2008;**70**:325-328

[66] Seesuriyachan P, Kuntiya A, Hanmoungjai P, Techapun C, Chaiyaso T, Leksawasdi N. Optimization of exopolysaccharide overproduction by *Lactobacillus confusus* in solid state fermentation under high salinity stress. Bioscience, Biotechnology, and Biochemistry. 2012;**76**(5):912-917

[67] Hussein M-D, Ghaly M, Osman M, et al. Production and prebiotic activity of exopolysaccharides derived from some probiotics. Egyptian Pharmaceutical Journal. 2015;**14**:1-9

[68] Korakli M, Pavlovic M, Vogel R. Exopolysaccharide and kestose production by *Lactobacillus sanfranciscensis* LTH2590. Applied and Environmental Microbiology. 2003;**69**:2073-2079

[69] Nguyen HT, Truong DH, Kouhoundé S, Ly S, Razafindralambo H, Delvigne F. Biochemical engineering approaches for increasing viability and functionality of probiotic bacteria. International Journal of Molecular Sciences. 2016;**17**(6):867

[70] Tada S, Katakura Y, Ninomiya K, Shioya S. Fed-batch coculture of *Lactobacillus* kefiranofaciens with Saccharomyces cerevisiae for effective production of kefiran. Journal of Bioscience and Bioengineering. 2007;**103**(6):557-562

[71] Bertsch A, Roy D, LaPointe G. Enhanced exopolysaccharide production by *Lactobacillus rhamnosus* in co-culture with *Saccharomyces cerevisiae*. Applied Sciences. 2019;**9**:4026

[72] Yamasaki-Yashiki S, Sawada H, Kino-Oka M, Katakura Y. Analysis of gene expression profiles of *Lactobacillus paracasei* induced by direct contact with *Saccharomyces cerevisiae* through recognition of yeast mannan. Bioscience of Microbiota, Food and Health. 2017;**36**(1):17-25

[73] Zeidan AA, Poulsen VK, Janzan J, et al. Polysaccharide production by lactic acid bacteria: from genes to industrial applications. FEMS Microbiology Reviews. 2017;**41**:Si68-S200

[74] Deo D, Davray D, Kulkarni R. A diverse repertoire of exopolysaccharide biosynthesis gene clusters in *Lactobacillus* revealed by comparative analysisin 106 sequenced genomes. Microorganisms. 2019;**7**:41

[75] Schilling C, Badri A, Sieber V, et al. Metabolic engineering for production of functional polysaccharides. Current Opinion in Biotechnology. 2020;**66**:44-51

[76] Alhudhud M, Humphreys P, Laws A. Development of a growth medium suitable for exopolysaccharide production and structural characterization by Bifidobacteriumanimalisspp. Lactis AD011. The Journal of Microbiological Methods. 2014;**100**:93-98

[77] Earle RR, Ayalasomayajula LU, Loknadh G, Reddy KSRK, Ratna L. A review of natural polymers used in pharmaceutical dosage forms. International Journal of Science and Research Methodology Human. 2016;**3**(3):77-88

Chapter 8

Effect of Sodium Acetate and Trace Element (Se^{2+}, Zn^{2+}) on Exopolysaccharide Production by *Lactobacillus plantarum* and Promote Antioxidant Capacity

Kusmiati Kusmiati, Yeni Yuliani, I. Nyoman Sumerta, Diah Radini Noerdjito, Wahidin Wahidin, Ghina Puspita Anggraeni, Yosephin Yosephin, Agung Tri Laksono and Atit Kanti

Abstract

L. plantarum producing EPS plays an important role as an antioxidant, antiproliferative, and anticancer. This study aims to increase the production of EPS by *L. plantarum* through modification of MRS (de Mann Rogosa Sharpe) media mixed with coconut water, treated with natrium acetate, Se, and Zn at different concentration, as well as understanding its effect on antioxidant activity. The effect of adding sodium acetate with different concentrations of 0.25, 0.50, 0.75, and 1.0% into mixed media MRS coconut water, (1:3) was studied. Fermentation experiments at different of Se^{2+} concentration (mM): 50; 75; 100; 125; 150; and 175, and addition of variation Zn^{2+} concentration (mM): 2.5; 5.0; 7.5; 10.0; 12.5; and 15.0), were carried out separately. Antioxidant potential was tested by FRAP (ferric reducing antioxidant power) and ABTS (2.2′-azinobis (3-ethyl benzatiazoline)-6-sulfonate). The results showed that the addition of sodium acetate with different concentrations showed a significant difference to the dry weight of EPS ($P < 0.05$). The increase in sodium acetate concentration was up to 1%, in line with the increase in EPS production by *L. plantarum* (g/g DW biomass). The addition of Se^{2+} 100 mM increased the ratio of glucose to protein content by 2.121. The value of the antioxidant activity of Fe (II) was 311.54, and the ABTS test obtained IC50 of 83.041. A separate experiment with the addition of Zn^{2+} in the fermentation medium of *L. plantarum* produced a fluctuating exopolysaccharide. The value of the antioxidant activity of Fe (II) M using the FRAP method was 275.886. The IC50 value with the ABTS method is 73.2942. Characterization of EPS from *L. plantarum* using FTIR (Fourier transforms infrared spectrophotometry) has hydroxyl, carboxylate, and aromatic functional groups.

IntechOpen

Keywords: exopolysaccharides, *Lactobacillus plantarum*, sodium acetate, Se^{2+}, Zn^{2+}, antioxidant

1. Introduction

Lactic acid bacteria (LAB) are widely used in the manufacture of traditional fermented milk. In the dairy industry, the bacteria are also useful as a culture starter in the fermentation process. Several strains of lactic acid bacteria have important roles in health which are beneficial microflora in the intestinal tract [1–3] and capable of synthesizing exopolysaccharides (EPS) [4–7]. Exopolysaccharides produced by lactic acid bacteria have been given increasing attention in recent years; it is due to their contribution to the rheology and texture of food products. In addition, EPS products, which are GRAS (generally recognized as safe), are declared safe for consumption and stable during storage [4, 5].

EPS is a polymer of high-molecular weight-reducing sugars, which are secreted by microorganisms into their external environment. This polymer has bioactivation, so it can be used for anti-viral and anti-inflammatory treatment. It has an inhibitory effect on tumor growth in vitro or in vivo [8–10]. The structure and composition of EPS is closely related to its anti-tumor biological function. In food industry, exopolysaccharides can function as thickeners, gelling agents, and emulsifiers. EPS from lactic acid bacteria (LAB) can exert functional effects on food, improve the rheology of fermented dairy products, and have beneficial health effects.

Several types of LAB that produce exopolysaccharides are *Lactobacillus acidophilus, L. rhamnosus, L. casei, L. reuteri, Bifidobacterium longum*, and *L. plantarum* [11]. The amount of exopolysaccharide produced by lactic acid bacteria is influenced by several factors, such as media composition (C, N concentration, and mineral supplementation), fermentation conditions, interactions between strains (co-culture fermentation), and fermentation technology (fed-batch fermentation). Other factors include physico-chemical conditions such as temperature, pH, level of oxygen presence, incubation time, and genetic factors [12].

Microbes need minerals to synthesize cellular components, to produce energy, and becomes electron acceptors in the metabolism of glucose and other sugars. Some minerals are enzyme activators for microbial metabolism such as Se^{2+}, Zn^{2+}, Mn^{2+}, Mg^{2+}, Ca^{2+}, and others. The addition of minerals with the right concentration into the growth medium of *L. plantarum* will increase the formation of exopolysaccharides.

Previous research on the effect of mineral species on EPS production and growth characteristics of *L. bulgaricus* strain ropy in milk media showed that the best mineral source was 0.5% sodium acetate, which yielded up to 476.6 mg/L of EPS compared to triammonium citrate, potassium phosphate, and magnesium sulfate at concentrations of 0–0.5% [13]. Referring to the results, this study added the mineral sodium acetate with a concentration variation of 0.25–1.0% into coconut water media to produce EPS by *L. plantarum*. The results of previous studies using coconut water producing the highest EPS of 7.0510 g with a composition of 75% coconut water [14].

The addition of selenium (Se^{2+}) micronutrients to MRS media was intended to review its effect on increasing antioxidant potential. Selenium as a microelement acted as a component of the enzyme glutathione peroxidase (GPX) which had antioxidant activity by reducing peroxide compounds, so it reduced free radicals in the body. Selenium is an essential function in the biological system [15]. The results of previous studies reported that the amount of selenium that could be added to bacterial culture was around 100–150 mM [16]. Micronutrient selenium became an important nutrient for cell proliferation that played its role in increasing cell growth [17]. Some LAB strains have been reported to be capable of resisting selenium oxyanions at high concentrations during cultivation. Especially, *L. plantarum* has been suggested as Se-enriched lactobacilli for food applications.

Research on micronutrients Zn and Cu added to goat's milk production showed an effect of increasing antioxidant activity. *L. casei* KCTC 3260, was found to possess a high antioxidant ability by chelating Fe^{2+} or Cu^{2+}, although no detectable SOD activity was observed [18]. This paper reports the results of research on the effect of adding selenium to the growth medium of *L. plantarum* bacteria for the production of exopolysaccharides that have potential as antioxidants.

Another micromineral is zinc (Zn) which is important for health. Zn is needed by various organs of the body, such as the skin of the gastrointestinal mucosa. Zn increased the antioxidant capacity of SOD which played its role in protecting pancreatic beta cells from damage caused by reactive oxygen species (ROS). Superoxide is one of the most abundant ROS produced by the mitochondria, while SOD catalyzes the breakdown of superoxide into hydrogen peroxide and water and is therefore a central regulator of ROS levels [19]. The complex form of copper zinc SOD (Cu Zn-SOD) compound was able to increase the activity of SOD. Study on the *L. fermentum* E-3 and E-18 were able to express Mn-SOD to resist oxidative stress [20]. Increased antioxidant activity of EPS from *L. plantarum* culture with the addition of zinc (Zn) with various concentrations of 2.5–15 mM will be reported here. Previous researchers reported that the uptake of zinc (Zn) in lactic acid bacteria (LAB) was 10 mM [13]. The results were used as a reference. Few studies about Zn-enriched LAB have been conducted, but it has been found that the bacterial growth and probiotic effect of *L. plantarum* can be enhanced by zinc in the gut [21].

In the past few years, intensive research of EPS from microbes has developed and widely applied as ingredients of functional foods, pharmaceuticals, nutraceuticals, cosmetics, and insecticides. EPS from microbes are predicted to quickly develop as fast as EPS from plants and microalga which now dominate the market. This is due to their role as an immunostimulatory [22, 23], antivirus activity, antibacterial, and anticancer [23–26]. The potential of EPS as antioxidant becomes crucial in the field of medicine and food industry due to their action as scavengers of reactive oxygen species (ROS). ROS are a diverse group of unstable and highly reactive oxygen-derived molecules, such as hydrogen peroxide (H_2O_2), hydroxyl radical (•OH), singlet oxygen ($^1O_2$), and superoxide (O_2-). These oxidants are produced under stress conditions that cause destruction of macromolecules (lipids, proteins, and DNA) and disrupt various redox signalling pathways in eukaryotic cells. The oxidative stress condition caused by the abnormally high levels of ROS triggers cardiovascular disease, and has also been implicated in diabetes, various types of cancer, neurologic and inflammatory diseases, ageing [27–29], and autoimmunological disorders [30].

Sustainable Development Goal 3 of the 2030 Agenda for Sustainable Development is to "ensure healthy lives and promoting well-being for all at all ages" [31]. One of the objectives that must be achieved is to reduce the mortality rate caused by non-communicable disease such as cardiovascular, cancer, diabetes, or chronic respiratory disease. In 2019, it was reported that globally, 74% of all mortality were caused by non-communicable diseases. In this regard, this study aims to produce EPS from *L. plantarum* bacteria as raw material for drugs that are useful for human health. The study focused on increasing the production of EPS through modification of MRS media mixed with coconut water by adding natrium acetate, Se, and Zn^{2+} and evaluating the potential of EPS as an antioxidant.

2. Materials and methods

2.1 Microorganism

Stock cultures of *L. plantarum* strains were preserved in a freezer at −80°C, in cryotubes containing 10% glycerol. The culture was stored at the Indonesian Institute of Sciences, which is located in Bogor, Indonesia. The strain was inoculated into MRS broth medium (Merck, Germany) for regeneration and incubated at 37°C.

2.2 Regeneration medium of *L. plantarum*

The regeneration medium used was solid MRS with the composition: MRS 5.22% w/v and agar 2% w/v. The materials were dissolved with distilled water up to 100 ml. Each was distributed as much as 4 ml into a tube and then sterilized by autoclaving at 121°C for 15 minutes with a pressure of 1 atm. The sterile medium is cooled to solidify to form slant agar.

2.3 Preculture of *L. plantarum*

Regenerated fresh *L. plantarum* bacteria were inoculated one to two loops into 25 ml of MRS broth medium; cultures were incubated at 37°C until they reached OD 1.0 as preculture.

2.4 Production of exopolysaccharides by *L. plantarum* bacteria

An amount of 2% precultured *L. plantarum* in logarithmic growth phase (OD 650 nm = ±1) was inoculated into a fermentation medium containing different microminerals to produce exopolysaccharides. Incubation was carried out at 37°C for 72 hours on a shaker incubator [32].

- MRS medium contained sodium acetate with different concentrations. The medium of fermentation used MRS 5.22% (w/v), it was dissolved in a mixture of distilled water and coconut water (with a ratio of 25:75) with the addition of sodium acetate with different concentrations, that is, 0.0; 0.25; 0.5; 0.75; and 1%.

- MRS medium contained selenium with different concentrations. The addition of Se^{2+} micronutrients to the MRS medium was 5.22% in a mixture of distilled water and coconut water (with a ratio of 25:75) was carried out with variations in concentrations of 50, 75, 100, 125, 150, and 175 mM.

- MRS medium contained zinc with different concentrations [33]. The addition of micronutrient Zn^{2+} into 5.22% MRS medium in a mixture of distilled water and coconut water (with a ratio of 25:75) was carried out with various concentrations of 2.5, 5.0, 7.5, 10, 12.5, and 15 mM.

2.5 Extraction of exopolysaccharides from *L. plantarum*

MRS medium modified with the addition of micronutrients was inoculated with *L. plantarum* bacteria at a concentration of 2% (v/v) and was cultivated at 37°C for 72 hours. Cultures of *L. plantarum* were boiled at 100°C for 15 minutes to inactivate enzymes that degraded exopolysaccharides. The cultures were cooled and centrifuged at 14.534×g, for 15 minutes, at 4°C, to separate the cell biomass and supernatant. The precipitated biomass was then dried and weighed as the weight of *L. plantarum* cell biomass (mg), while the supernatant was extracted to obtain EPSs. An amount of two times of the volume of 96% ethanol was added to the supernatant for precipitation of EPSs. The mixture was stored at 4°C for 24 hours. Then, the EPS was collected by centrifugation (11.772×g, 4°C, 20 minutes). The pellets were dissolved with distilled water. EPSs were purified by adding 15% (w/v) trichloroacetic acid (TCA) [34], then was centrifuged (20,000×g, 10 minutes, 4°C). The pellets were dried at 55°C, weighed as EPS dry weight (mg) [34].

2.6 Total sugar analysis

Total sugar content was determined by the modified phenol-sulfuric acid method using glucose as a standard [35]. An amount of 1 ml of the sample was mixed with 0.5 ml of 5% phenol solution. Then, 2.5 ml of 95% sulfuric acid was added. The sample were incubated at 25°C for 10 minutes; then stirred for 1 minute. The sample solution was re-incubated for 20 minutes at 25°C. The absorbance of each sample was measured using a spectrophotometer at a wavelength of = 490 nm. Sample control used distilled water (1 ml).

2.7 Protein analysis by the method of Lowry

The standard solution used in protein analysis was bovine serum albumin (BSA). An amount of 0.5 ml of the sample was added with 0.5 ml of 1 N NaOH, shaken, and boiled at 100°C for 20 minutes, then cooled. An amount of 0.5 ml of the mixed solution (5% Na_2CO_3, 1% $CuSO_4.5\ H_2O$ and 2% Na-K-tartrate) was added to each sample; and was shaken homogeneously. Add 0.5 ml of Folin-Ciocalteu reagent to each sample; and was shaken homogeneously and let stand for 30 minutes. The absorbance of the BSA standard solution and the test sample was measured with a UV-Vis spectrophotometer at a wavelength of 750 nm. Sample control used distilled water [36].

2.8 Functional groups of EPS analysis by Fourier-transform infrared (FT-IR) spectroscopy

Sample were prepared by grinding dried EPS (1 mg) with KBr (20 mg) in 1:20 w/w ratio and pressed into 1 mm thick pellets for measurement. FT-IR spectra were recorded on an FT-IR spectrometer (Shimadzu, Japan) in the frequency range of 4000–400 cm^{-1} [37].

2.9 In vitro antioxidant activities of EPS

2.9.1 Ferric reducing antioxidant power measurement of EPS

Various concentrations of exopolysaccharide (0, 50, 75, 100, and 125 ppm) was blended with 1 ml of 0.2 M phosphate buffer, pH = 6.6 and 1 ml of 1% potassium ferricyanide solution. The mixture incubated at 50 °C for 20 minutes. Then 1 ml of trichloroacetic acid (10%) was added to the reaction mixture and centrifuged at 3000 rpm for 10 minutes. Then 2.5 ml of the supernatant was mixed with distilled water (2.5 ml) and 0.5 ml of 0.1% ferric chloride solution. The absorbance of the solution was measured at 700 nm using a UV spectrophotometer. Increasing the absorption of the samples means increasing the reducing power of the samples. The blank sample was distilled water, and ascorbic acid was used as positive reference standard. The ferric ion reducing power were expressed as milligrams of ascorbic acid equivalent (AAE) per ml of EPS sample [38, 39].

2.9.2 Assay of ABTS+ radical scavenging capacity

The antioxidant activity of EPS was assay based on the radical scavenging capacity of 2,2′-azinobis-(3-ethylbenzothiazoline-6-sulfonic acid (ABTS+) radical with slight modification. The maximum absorption wavelength of ABTS radical ions is 734 nm, and hence the absorbance at this particular wavelength is used to detect the concentration of ABTS radical ions. Briefly, ABTS solution (7.4 mM) was mixed with potassium persulphate solution (2.6 mM) (1:1 v/v) and was left at 20–22°C in the dark for 24 hours. The ABTS stock solution was diluted with absolute ethanol to an absorbance of 0.7 ± 0.02 at 734 nm. Then, 0.2 and 0.8 ml of ABTS working solution were mixed thoroughly. VC was used as the positive control. The absorbance of the mixture solution was determined at 734 nm after 5 minutes of incubation in the dark, using a microplate reader. The ABTS radical scavenging activity (%) was calculated using the following formula:

$$\text{Scavenging rate}\,(\%) = (\text{Ac} - \text{As})\,/\,\text{Ac} \times 100 \quad (1)$$

Whereas is the absorbance of the test sample and Ac is the absorbance of the control at 734 nm [40–42].

2.10 Data analysis

Data were analyzed by one way ANOVA with three replications using SPSS ver 22.0 with $P = 0.05$. This analysis was then followed by Duncan multiple range test.

3. Results and discussion

3.1 Effect of sodium acetate concentration on the formation of EPS *L. plantarum*

Exopolysaccharide was extracted from the culture of lactic acid bacteria *Lactobacillus plantarum* fermented in coconut water-MRS medium in a ratio (25:75) at 37°C for 72 hours. The use of coconut water with a percentage of 75% referred to previous studies because it was the best percentage to produce the exopolysaccharide *L. plantarum* [14]. Liquid MRS medium was used because it was a selective medium for the growth of lactic acid bacteria. Seesuriyachan's research used MRS media which was a good growth medium for lactic acid bacteria; the addition of coconut water and sucrose as a carbon source could affect the yield of exopolysaccharides. The research resulted in the production of exopolysaccharides which continued to increase, that is, from a concentration of 0–100% coconut water also contains protein, sugar, amino acids, various vitamins and minerals, so with the relatively complete nutritional content, coconut water is very potential to be used as a basic material for fermenting organic acids and as an alternative carbon source for the production of exopolysaccharides from lactic acid bacteria. Lactic acid-producing bacterium is one of the

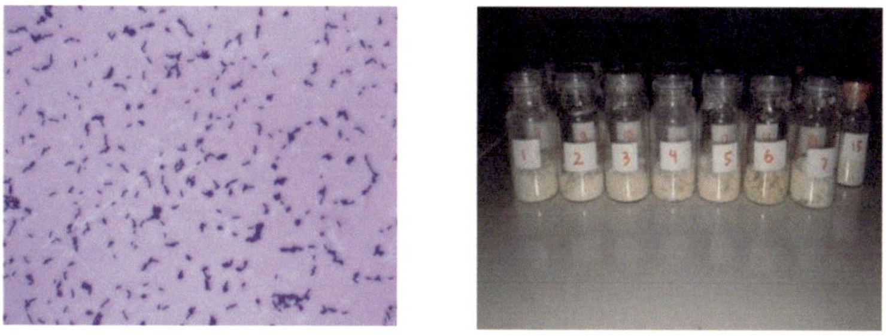

(a) (b)

Figure 1.
(a) Morphology of whole cells of L. plantarum *visualized by microscop 1000×. (b) EPS crude extracted from* L. plantarum.

bacteria that tends to be attracted to sugar-containing habitats such as coconut water [43]. **Figure 1** shows the cell morphology of *L. plantarum* with methylene blue dye and crude exopolysaccharide extracted from *L. plantarum* culture.

The fermentation temperature used was 37°C, which was the optimal temperature for the growth of lactic acid bacteria. The fermentation time for 72 hours was the optimal time for harvesting exopolysaccharides for the growth of lactic acid bacteria. The supernatant obtained from the extraction was then added with cold 96% ethanol in a ratio of 1:1; then allowed to stand for 48 hours at 4°C. The addition of cold 96% ethanol aimed to precipitate exopolysaccharides [32].

The amount of exopolysaccharide produced by lactic acid bacteria was influenced by several factors such as media composition (source and concentration of carbon and nitrogen), fermentation conditions, effects of growth media (mineral supplementation), interactions between strains (co-culture fermentation), and pharmaceutical technology (fed-batch fermentation), as well as physico-chemical conditions of bacterial growth such as temperature, pH, level of oxygen presence, incubation time, and genetic factors [12].

Table 1 shows that the production of EPS (g) per dry cell biomass (g) was increasing, in line with the addition of sodium acetate concentration in MRS medium: coconut water (75:25). A significant increase in EPS production against the control (media without sodium acetate) began to be seen at a concentration of 0.75% sodium acetate: increased to 23.81%. At 1% sodium acetate concentration, EPS production increased to 34.94%. The greater the concentration of sodium acetate used the more exopolysaccharide compounds produced. It was influenced by the addition of sodium

No.	Na. acetate concentration (%)	Production of EPS/cell biomass dry (g/g)
1	0	28.22^{a}
2	0.25	29.47^{a}
3	0.5	31.75^{ab}
4	0.75	34.94^{bc}
5	1.0	38.09^{c}

Table 1.
Dry weight of EPS L. plantarum *on MRS Media: Coconut water (25:75) with variations in sodium acetate concentration. The numbers followed by the same letter are not significantly different ($P < 0.05$).*

acetate mineral as an electron acceptor in the metabolism of glucose and other sugars which helped lactic acid bacteria to produce exopolysaccharide compounds [44].

In the exopolysaccharide polymerization reaction, the formation of carbon chains required minerals as electron acceptors binding one monomer to others. Lactic acid bacterium was not limited to oxygen as an electron acceptor. Anaerobically, some organic components could be treated with the same purpose as electron acceptors. This case particularly happened at obligate heterofermentative LAB in the alcohol or acetate formation pathway. However, it turned out that organic electron acceptors could play an important role in homofermentative LAB in anaerobic metabolism on certain substrates [44].

Research by Pham et al. also used several types of minerals in the medium to see the activity of *L. rhamnoses* cells in producing exopolysaccharides. The results showed that during the initial exponential growth phase the biosynthesis of exopolysaccharides did not occur. Production occurred in a stationary phase leading to death and then the exopolysaccharide produced could be reused by microbes as a carbon source due to the presence of enzymes produced by the bacteria themselves that could degrade exopolysaccharides. Consequently, the prolongation of the incubation time decreased the production of exopolysaccharides [45].

The extraction method used could also affect the amount of exopolysaccharide produced. The heating step carried out at the beginning of the extraction was able to increase the recovery of exopolysaccharides, then increased the number of exopolysaccharides obtained. The addition of ammonium sulphate at the time of extraction aimed to precipitate proteins and separate exopolysaccharide compounds from a mixture of other compounds [45].

The production of exopolysaccharides from microbial cultures depended on several different parameters. The formation of polysaccharides was most often associated with the presence of carbohydrates and low or high temperatures. Mineral requirement was also a critical factor. Restriction of nitrogen, phosphorus or sulphur sources increased the production of exopolysaccharides; on the other hand, some researchers reported that phosphate and trace elements were essential elements for the synthesis of exopolysaccharides by *Pseudomonas aeruginosa.* Several types of minerals were needed by microbes as growth factors needed to form energy and compose cell components and the formation of secondary metabolites. Micro elements such as K, Ca, Mg, Cl, Fe, Mn, Co, Cu, Zn, and Mo were needed by almost all microbes. The transfer of nutrients into microbial cells could be in the form of passive diffusion, diffusion with the help of permease, active transfer or through the phosphotransferase system. Minerals were generally transferred by active transfer [44].

3.2 Effect of sodium acetate concentration on total sugar content and protein in EPS

Determination of glucose levels in exopolysaccharide samples was carried out using the phenol-sulphate method using visible light spectrophotometry. This method was chosen to determine the total sugar content in exopolysaccharides because it was commonly used to determine the total carbohydrate content of bacterial polysaccharides. The advantage of this method was that the reagents used were cheap and easy to obtain, little equipment was needed and the analysis was simple [46]. The principle of this method was that simple sugar and oligosaccharide could react with phenol in concentrated sulfuric acid to produce a stable yellowish orange color. The addition of phenol and concentrated sulfuric acid aimed to form a color complex in the sample so

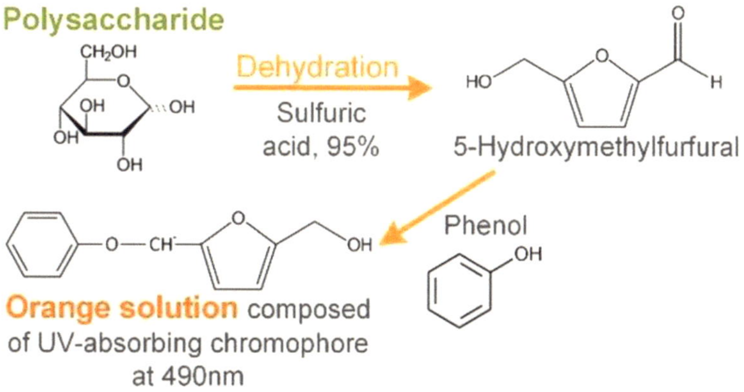

Figure 2.
Schematic overview of the phenol sulfuric method for the total carbohydrate analysis [47].

that it could be detected by UV-VIS spectrophotometry. The addition of concentrated sulfuric acid would produce a yellowish orange hydroxymethyl furfural compound absorbing at a wavelength of 490 nm [47]. The mechanism of the dehydration reaction of glucose to hydroxy methyl furfural can be seen in **Figure 2**.

Figure 3a: the correlation coefficient value was 0.9924 showing that there was a linear relationship between the concentration of the standard glucose comparison solution and the absorption at 490 nm; the increase in concentration was proportional to the increase in absorption. **Figure 3b**: analysis of variance (ANOVA) and DMRT test (Duncan multiple range test) showed that there was a significant difference in the addition of sodium acetate with a concentration of 0–0.75% with an increase in glucose levels at that concentration. Meanwhile, at the concentration of sodium acetate concentration of 0.75% with 1% there was no significant difference even though glucose levels in the 1% sodium acetate treatment increased. **Figure 3b** the highest glucose level was obtained in 1% sodium acetate sample (75.54%) and the lowest was found in 0% sodium acetate sample (29.05%).

The absorption of BSA standard solution with concentrations of 20, 40, 60, 80, and 100 ppm was measured by a spectrophotometer at a wavelength of 750 nm (39)

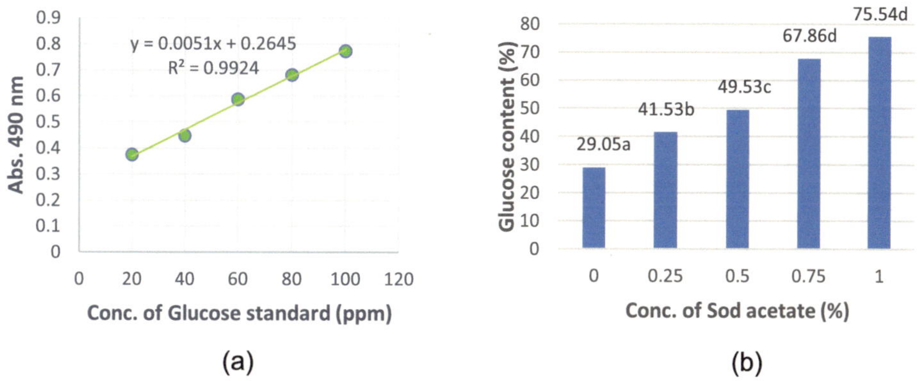

Figure 3.
(a) Relationship of standard glucose concentration to absorption at λ490 nm and (b) effect of Na acetate concentration on glucose levels in EPS (%). The numbers followed by the same letter are not significantly different ($P < 0.05$).

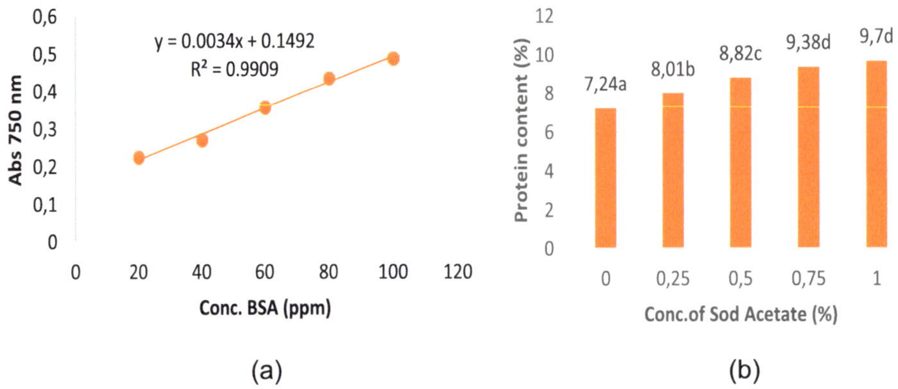

Figure 4.
(a) Relationship of protein standard concentration (BSA) to absorption at λ750 nm and (b) effect of Na acetate concentration on protein content in EPS (%). The numbers followed by the same letter are not significantly different (P < 0.05).

resulting in a calibration curve with the regression line equation y = 0.0034x + 0.1492 with the correlation coefficient value obtained 0.9909 (**Figure 4a**).

Determination of protein content using the Lowry method based on two different reactions. The first reaction was the formation of copper Cu+. Under alkaline conditions formed by a solution of Na_2CO_3 in NaOH, Cu^{2+} ions formed a complex with peptide bonds reducing Cu^{2+} to Cu+. The second reaction was a reduction reaction by Folin-Ciocalteu reagent (phosphomolybdate and phosphotungstate). The Cu+ ion and the radical groups of tyrosine and tryptophan reacted with Folin's reagent to produce an unstable product that reduces molybdenum or tungsten blue. Protein would react with Folin-Ciocalteu reagent to form a complex compound giving a blue color [48].

The results of analysis of variance (ANOVA) and DMRT showed a significant increase in protein content due to treatment with sodium acetate concentration (0–0.75%). However, the results did not show a significant difference between 0.75 and 1% sodium acetate concentrations. **Figure 4b** shows that the highest protein content was obtained as a result of 1% sodium acetate treatment (9.70%) and the lowest was found in 0% sodium acetate treatment (7.24%).

3.3 Interpretation of IR spectrum of EPS and glucose samples

The functional groups of exopolysaccharide samples produced from *L. plantarum* isolates with comparison, namely glucose, were determined using an FT-IR (Fourier transform infra-red) spectrophotometer [49].

The results of the interpretation of the infrared spectrum of the exopolysaccharide sample of *L. plantarum* and the reference standard (glucose) are presented in **Table 2**. Glucose was used as a comparison because glucose is the main compound contained in the exopolysaccharide.

The results of the FT-IR analysis in **Table 2** show that the exopolysaccharide and glucose samples contained IR spectra similarities indicating a typical absorption of polysaccharide compounds. Glucose showed the presence of hydroxyl groups (-OH), alkane groups (–CH), and C–C vibrations. Meanwhile, the exopolysaccharide isolated from *Lactobacillus plantarum* showed the presence of a hydroxyl group (-OH), a carbonyl group (–C=O) in the carboxylate and a C–C vibration. These results indicated that the extract obtained was a carbohydrate compound.

Sample code	Wave number (cm^{-1})	Absorption peak sample (cm^{-1})	Function groups
Glucose	3500–3200	3275.13	Group –OH (hydroxyl)
3000–2900	2935.66	*Stretching* C-H (alkanes)	
1300–1000	1224.80, 1151.50, 1109.07	Vibration C–C and C-O	
EPS	3500–3200	3219.19	Group –OH (hydroxyl)
1800–1600	1678.07	Group –C=O (carboxylate)	
1600–1475	1539.20	Group C=C (aromatic)	
1300–1000	1085.92	Vibration C–C and C-O	

Table 2.
Interpretation of IR Spectrum of Lactobacillus plantarum *EPS samples and glucose standard.*

The area of 200–900 cm^{-1} was the fingerprint area for carbohydrates and could be used for carbohydrate identification. Exopolysaccharides from *L. plantarum* with the addition of sodium acetate had an absorption band of 1085 cm^{-1}. Absorption bands in the area around 1080 cm^{-1} were characteristic for carbohydrates derived from microbial biomass glucose, galactose, and mannose had absorption bands in the region of 983–1200 cm^{-1} [50]. The sample and glucose had an absorption band in that area. Specifically, glucose was in the wave number of 1100–1124 cm^{-1} and the exopolysaccharide sample was in the wave number of 1000–1150 cm^{-1}. It indicated the presence of sugar monomers such as glucose, galactose and mannose in the sample. Thus, the sample can be said to be a polysaccharide based on the resulting IR spectra and its similarity to glucose.

3.4 Effects concentration variations of Se^{2+} and Zn- in fermentation media on dry weight of cell biomass (g), EPS production (g), glucose, and protein levels (%)

Table 3 shows that the higher the concentration of Se in the medium, the higher the cell biomass formed. It was inversely proportional to the formation of EPS: the higher the selenium concentration in the medium, the lower the dry weight of EPS. The decrease in EPS yield was due to the increase in selenium concentration in the media because the selenium inhibited the metabolic process of EPS synthesis by *L. plantarum* [51].

The results of Arain *et al.* reported that the addition of mineral treatment could change the pH of the media which caused the inhibition of the growth of *Lactobacillus bulgaricus* due to the high concentration of minerals. Mineral requirements for microbial growth were only in small amounts. This statement supported that the concentration of selenium as a mineral when added at high concentrations can result in inhibition of EPS production. Selenium is essential for the progression and expression of humoral and cell-mediated immune responses. Selenium enhances the phagocytic activity of neutrophil granulocytes and macrophages, and when stimulated in myogens, increases T lymphocyte counts [52].

Zinc is involved in protein synthesis and is a regulator of involved in protein synthesis an is a regulator of synaptic activity and neuronal processes. Zinc deficiency due to the reduced ingestion of this metal causes an abnormal process in the nervous system [51]. The effect of adding Zn^{2+} to medium with different concentrations is shown in **Table 4**: the higher the zinc concentration in the growth medium, the higher the cell biomass produced. On the other hand, the production

No.	Treatment Conc. of Se^{2+} (mmol)	Cell biomass dry weight (g)	Crude EPS production (g) per 25 ml medium	Glucose content in EPS (%)	Protein content in EPS (%)	Ratio of glucose: Protein
1.	0	$0.023^{a} \pm 0.002$	$2.781^{g} \pm 0.046$	$18.435^{a} \pm 0.883$	$14.172^{a} \pm 0.483$	1.308
2.	50	$0.041^{b} \pm 0.003$	$2.631^{f} \pm 0.022$	$35.200^{b} \pm 0.975$	$16.916^{ab} \pm 1.293$	2.091
3.	75	$0.045^{bc} \pm 0.003$	$2.558^{e} \pm 0.014$	$36.826^{b} \pm 1.430$	$18.772^{b} \pm 2.075$	2.018
4.	100	$0.052^{bcd} \pm 0.001$	$2.433^{d} \pm 0.021$	$42.849^{c} \pm 1.814$	$20.229^{b} \pm 1.429$	2.121
5.	125	$0.057^{cd} \pm 0.001$	$2.343^{c} \pm 0.037$	$48.345^{d} \pm 1.729$	$24.687^{c} \pm 3.455$	1.980
6.	150	$0.063^{d} \pm 0.004$	$2.246^{b} \pm 0.021$	$48.070^{d} \pm 2.138$	$32.380^{d} \pm 1.448$	1.488
7.	175	$0.077^{e} \pm 0.002$	$2.165^{a} \pm 0.026$	$55.169^{e} \pm 2.522$	$35.047^{d} \pm 2.061$	1.574

Data are means ± S.D. of three replicates.

Table 3.
Effects of Se^{2+} treatment with different concentrations on dry weight of cell biomass (g), EPS production (g), glucose, and protein levels (%). The numbers followed by the same letter are not significantly different ($P < 0.05$).

No.	Treatment of Zn^{2+} (mmol) conc.	Cell biomass dry weight (g)	Crude EPS production (g)	Glucose content in EPS (%)	Protein content in EPS (%)	Ratio of glucose: Protein
1.	0	$0.0148^{a} \pm 0.0004$	$3.9539^{a} \pm 0.063$	$18.97^{a} \pm 0.551$	$12.57^{a} \pm 0.551$	1.510
2.	2.5	$0.0144^{a} \pm 0.0005$	$3.6430^{b} \pm 0.094$	$35.70^{b} \pm 1.015$	$27.63^{c} \pm 1.002$	1.293
3.	5.0	$0.0149^{a} \pm 0.0004$	$3.8763^{ab} \pm 0.042$	$46.57^{c} \pm 0.873$	$38.85^{e} \pm 1.626$	1.203
4.	7.5	$0.0176^{b} \pm 0.0006$	$3.7469^{ab} \pm 0.084$	$50.87^{d} \pm 0.503$	$27.73^{c} \pm 1.150$	1.836
5.	10	$0.0190^{c} \pm 0.0007$	$2.9172^{c} \pm 0.254$	$69.20^{e} \pm 1.082$	$19.03^{b} \pm 1.193$	3.650
6.	12.5	$0.0226^{d} \pm 0.0004$	$2.7365^{c} \pm 0.211$	$50.73^{d} \pm 0.986$	$10.73^{a} \pm 1.193$	4.760
7.	15	$0.0324^{e} \pm 0.0003$	$2.4257^{d} \pm 0.225$	$47.93^{e} \pm 0.665$	$30.83^{d} \pm .250$	1.55

Data are means ± S.D. of three replicates.

Table 4.
Effects of Zn^{2+} treatment with different concentrations on dry weight of cell biomass (g), EPS production (g), glucose and protein levels (%). The numbers followed by the same letter are not significantly different ($P < 0.05$).

of exopolysaccharides tended to decrease. The inversely proportional results of cell biomass and exopolysaccharides had been carried out in a study. The yield of exopolysaccharide production was inversely proportional to cell biomass because of the need for energy support in the exopolysaccharide biosynthesis process. The limitation of energy production in cells during the formation of exopolysaccharides caused the amount of biomass to be low, because the formation of the two was interrelated. Higher exopolysaccharides were produced with lower cell growth. It was possible that the cell growth pathway was blocked and converted to an exopolysaccharide synthesis pathway [43].

The production of exopolysaccharides obtained from lactic acid bacteria was influenced by several factors such as fermentation conditions, effects of growth media (mineral supplementation), interactions between strains (co-culture fermentation), and fermentation technology (fed-batch fermentation). Regulation of cell growth at a constant pH resulted in better exopolysaccharide yields. The acidification process occurred because the production of lactate caused the glycohydrolase enzyme to become active (pH range 5). It caused the yield of EPS to decrease due to the enzymatic process. The culture conditions and the composition of the media (not only carbon sources) affect the yield of EPS and the molecular characteristics of the biopolymer [53].

Glucose equivalent EPS levels were determined using the phenol-sulfate method. Glucose did not have a chromophore so it was supposed to be reacted with phenol-sulphate to form an orange-yellow color having a maximum absorption at 490 nm. The addition of concentrated sulfuric acid H_2SO_4 caused hydrolysis of glycosidic bonds in EPS so that furfural compounds or furfural derivatives were formed which would condense with phenol to form yellow-orange compounds. The results of the measurement of glucose uptake and concentration in EPS can be seen in **Table 5**. The highest average glucose level was obtained in exopolysaccharide samples with the addition of Se^{2+} 175 mM concentration reaching 55.169% and the addition of 10 mM zinc of 69.20%, while the lowest glucose level was obtained at exopolysaccharide samples in the media without the addition of Se^{2+} and Zn^{2+}.

Glucose levels in exopolysaccharides would increase because during fermentation, lactose would be broken down into glucose and galactose which became the main carbon sources in increasing the activity of the UDP-glucose pyro phosphorylase and UDP galactose-4-epimerase enzymes. UDP galactose-4-epimerase was a key enzyme in the formation of EPS, and the enzyme would be active when there was a sugar unit in the form of galactose as a precursor in the formation of exopolysaccharides [53].

Analysis of protein content in EPS samples was determined using the Lowry method by visible light spectrophotometry. The protein in the sample would react with copper (II) sulphate under alkaline conditions to form Cu^{2+} ions and an amino acid radical group which then reacted with Folin-Ciocalteu to produce an unstable product that reduces molybdenum or tungsten blue. The reaction produced a blue-colored complex which gives the maximum absorption at 750 nm. Protein levels would decrease because the longer fermentation means the longer the opportunity for lactic acid bacteria to degrade protein, so it caused protein levels to decrease.

The results of the calculation of the ratio of high levels of glucose to protein provided information on the optimum culture conditions for producing EPS, because glucose was a monomer of EPS. The culture produced EPS with the highest glucose: protein ratio at the concentrations of Se^{2+} 100 mMol and Zn^{2+} 12.5 mMol.

No.	Conc. Se^{2+} (mmol)	Se^{2+} treatment		Conc. Zn^{2+} (mmol)	Zn^{2+} treatment	
		FRAP* (mg $FeSO_4$/mg)	ABTS** IC50 (mg/ml)		FRAP (mg $FeSO_4$/mg)	ABTS IC50 (mg/ml)
1.	0	0.223 ± 0.008	107.320	0	0.025 ± 0.001	117.806
2.	50	0.245 ± 0.006	146.930	2.5	0.091 ± 0.001	71.172
3.	75	0.273 ± 0.038	84.096	5.0	0.168 ± 0.004	76.993
4.	100	0.254 ± 0.030	76.944	7.5	0.331 ± 0.008	72.254
5.	125	0.276 ± 0.003	74.929	10	0.438 ± 0.029	69.122
6.	150	0.285 ± 0.005	80.660	12.5	0.471 ± 0.020	66.645
7.	175	0.289 ± 0.018	61.882	15	0.479 ± 0.035	55.535

**FRAP Data are means ± S.D. of two replicates.*
***The ABTS radical is completely reduced to this concentration accompanied by the disappearance of the green color.*

Table 5.
Antioxidant activity test of EPS L. plantarum *using FRAP and ABTS methods.*

3.5 Antioxidant activity test of *L. plantarum* exopolysaccharide with FRAP (ferric reducing antioxidant power) and ABTS (2,2′-azinobis (3-ethylbenzatiazoline)-6-sulfonate) methods

The results of the exopolysaccharide antioxidant activity test using the FRAP method showed that exopolysaccharides were potential to be antioxidants based on their ability to reduce colorless Fe3+ TPTZ (2,4,6-tripyridyl-s-triazine) to blue Fe^{2+}, so that the antioxidant power of a compound was analogous to reducing ability of the compound. Fe3+ TPTZ compounds represented oxidizing compounds that might be present in the body and could damage body cells, while samples contained antioxidants which could then reduce Fe3+ TPTZ compounds to Fe^{2+} TPTZ so that Fe3+ TPTZ compounds would not carry out reactions that damage body cells. The more the concentration of Fe3+ TPTZ reduced by the sample to Fe^{2+} TPTZ, the higher the antioxidant activity of the sample.

The reducing activity of the EPS samples was determined according to the method described by Benzie and Strain [54]. Calibration using $FeSO_4$ which expressed as mg Fe(II) per gram extract (**Figure 5a**). The antioxidant activity test using the ABTS method using vitamin E as a comparison (**Figure 5b**), showed an IC50 of 8.3090 (μg/ml). The value of antioxidant test results from EPS *L. plantarum* using the FRAP and ABTS methods are listed in **Table 5**.

The results of the antioxidant activity test of EPS *L. plantarum* treatment with a concentration of Se^{2+} 175 mmol showed the highest FRAP value, namely 0.289 ± 0.018 mg $FeSO_4$/mg; and the best antiradical ABTS activity, namely IC50 61,882 mg/ml. Meanwhile, Zn^{2+} 15 mmol treatment showed the highest FRAP value, namely 0.479 ± 0.035 mg $FeSO_4$/mg; and the best antiradical ABTS activity, namely IC50 55,535 mg/ml. The results showed that EPS *L. plantarum* had very strong antioxidant potential. The principle of the ABTS method was that EPS antiradical compounds would ward off free radicals marked by the loss of blue color (decolorization) in the ABTS reagent. It was indicated by a decrease in the absorbance value of the measured sample absorption. The advantages of ABTS compared to other methods were that the test was simple and easy to repeat; and the most important thing was that it was flexible and could be used to measure the activity of both hydrophilic and lipophilic antiradicals [55].

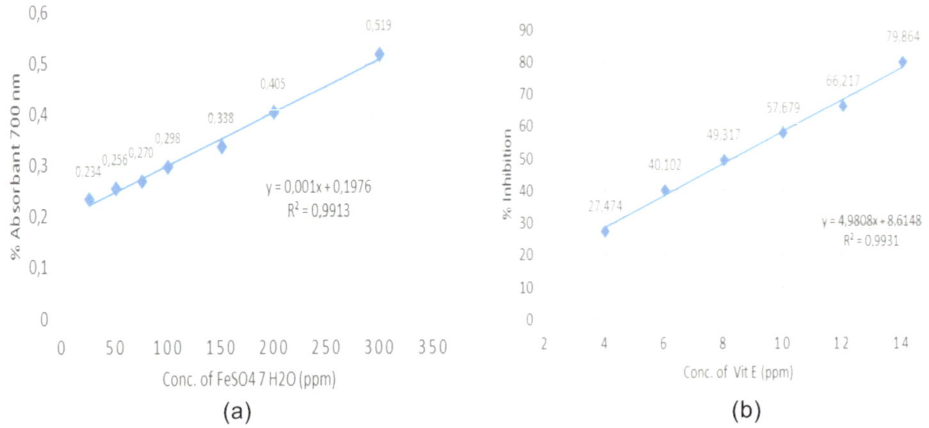

Figure 5.
(a) Standard curve of $FeSO_4.7H_2O$ and (b) relationship of vitamin E concentration (ppm) to % inhibition.

4. Conclusions

The production of EPS derived from the group of lactic acid bacteria including *L. plantarum* was in great demand because it was safe for consumption. The effect of EPS to improve food texture was a strategy to reduce the use of additives. The potential of EPS to be applied in the pharmaceutical field opened up opportunities for researchers to increase EPS production through modification of the fermentation medium by adding coconut water, Na-Acetate and minerals (Se^{2+} and Zn^{2+}). The increase in EPS production by *L. plantarum* occurred due to the addition of natrium acetate concentration of up to 1%. Natrium acetate concentration of 0.75% gave significantly different EPS results to the control. EPS production by *L. plantarum* was optimum at 100 mmol of Se^{2+} concentration and 12.5 mmol of Zn^{2+} concentration, based on the highest ratio of glucose: protein.

The effect of Se and Zn concentrations on the antioxidant activity of EPS using the FRAP and ABTS methods shows that the higher the concentration of these trace elements, the higher the antioxidant activity.

The addition would affect the increase in enzyme activity for the synthesis of EPS. The antioxidant activity test of the *L. plantarum* EPS compound showed a very strong category.

Acknowledgements

The authors are grateful to Dr. Arif Nurkanto, Head of Laboratory in Research Center for Biosystematics and Evolution, National Research and Innovation Agency (BRIN), for their valuable support to complete this study. We express our gratitude to Apt. Lisana Sidqi Aliya, S. Farm., M. Biomed., and Apt. Dr. Herdini, Lecturer at Fac. of Pharmacy-ISTN, Jakarta for suggestions during the research.

Conflict of interest

The authors declare that there are no competing interests.

Author details

Kusmiati Kusmiati[1]*, Yeni Yuliani[1], I. Nyoman Sumerta[1], Diah Radini Noerdjito[1], Wahidin Wahidin[2], Ghina Puspita Anggraeni[3], Yosephin Yosephin[3], Agung Tri Laksono[3] and Atit Kanti[1]

1 Research Center for Biosystematics and Evolution, National Research and Innovation Agency (BRIN), Cibinong, West Java, Indonesia

2 Faculty of Pharmacy, UTA'45 University, Jakarta, Indonesia

3 Faculty of Pharmacy, National Institute of Science and Technology, Jakarta, Indonesia

*Address all correspondence to: kusmiati@brin.go.id

References

[1] Behera SS, Ray RC, Zdolec N. *Lactobacillus plantarum* with functional properties: An approach to increase safety and shelf-life of fermented foods. Hindawi BioMed Research International. 2018;**18**:1-18

[2] Axelsson L, Ahrne S. Lactic acid bacteria. Applied Microbial Systematics. Netherlands: Springer; 2000. pp. 367-388

[3] Ray RC, Joshi VK. Fermented foods: Past, present and future scenario. In: Ray RC, Montet D, editors. Microorganisms and Fermentation of Traditional Foods. Boca Raton, FL: CRC Press; 2014. pp. 1-36

[4] Park SY, Ji HP, et al. Evaluation of functional properties of lactobacilli isolated from Korean white kimchi. Food Control. 2016;**69**:5-12

[5] Gharaei-Fathabad E, Eslamifar M. Isolation and applications of one strain of *Lactobacillus paraplantarum* from tea leaves (*Camellia sinensis*). American Journal of Food Technology. 2011;**6**:429-434. DOI: 10.3923/ajft.2011.429.434

[6] De Vuyst L, Degeest B. Heteropolysaccharides from lactic acid bacteria. FEMS Microbial Review. 1999;**23**(2):153-157

[7] Ricciardi A, Clementi F. Exopolysaccharides from lactic acid bacteria. Structure–production and technological applications. Italian Journal of Food Science (Italy). 2001;**12**(1):23-45

[8] Liamas I, Mata JA, Tallon R, Bressollier P, Maria C, Queseda E, et al. Characterization of the exopolysaccharide produced by *Salpiger mucosus* A3T, a halophilic species belonging to the Alphaproteobacteria, isolated on the Spanish Mediterranean Seaboard. Marine Drugs. 2010;**8**:2240-2251

[9] Sungur T, Aslim B, Karaaslan C, Aktas B. Anaerobe impact of exopolysaccharides (EPSs) of *Lactobacillus gasseri* strains isolated from human vagina on cervical tumor cells (HeLa). Anaerobe. 2017;**47**:137-144. DOI: 10.1016/j. anaerobe.2017.05.013

[10] Adesulu-dahunsi AT, Jeyaram K. Production of exopolysaccharide by strains of *Lactobacillus plantarum* YO175 and OF101 isolated from traditional fermented cereal. Beverage. Peer-Reviewed Journal. 2018:1-21. DOI 10.7717/peerj.5326

[11] Tallon R, Bressollier P, Urdaci MC. Isolation and characterization of two exopolysaccharides produced by *Lactobacillus plantarum* EP56. 2006. Available from: http://pasteur.fontismedia.com/infiles.doc

[12] Nwodo UU, Green E, Okoh AI. Bacterial exopolysaccharides: Functionality and prospects. International Journal of Molecular Sciences. 2012;**13**:14002-14015

[13] Leonardi A, Zanoni S, De Lucia M, Amaretti A, Raimondi S, Rossi M. Zinc Uptake by Lactic Acid Bacteria. 2013

[14] Stephany N. The effect of coconut water medium on the exopolysaccharides production using four isolates of lactic acid bacteria. Jakarta, Indonesia: Faculty of Pharmacy, National Institute of Science and Technology; 2017 (unpublished)

[15] Cremonini E, Zonaro E, Donini M, Lampis S, Boaretti M, Dusi S, et al. Biogenic selenium nanoparticles:

Characterization, antimicrobial activity and effects on human dendritic cells and fibroblasts. Biotechnology. 2016;**9**:758-771

[16] Shixue Z, Jing S, Liang W, Rong Y, Dan W, Yujia D, et al. Selenite reduction by the obligate aerobic bacterium *Comamonas testosteroni* S44 isolated from a metal contaminated soil. BMC Microbiology. 2014;**14**:204

[17] Huawei Z. Selenium as an essential micronutrient: Roles in cell cycle and apoptosis. Molecules. 2009;**14**(3):1263-1278

[18] Lee J, Hwang KT, Chung MY, Cho D, Park C. Resistance of *Lactobacillus casei* KCTC 3260 to reactive oxygen species (ROS): Role for a metal ion chelating effect. Journal of Food Science. 2005;**70**:m388-m391

[19] Landis GN, Tower J. Superoxide dismutase evolution and life span regulation. Mechanisms of Ageing and Development. 2005;**126**:365-379

[20] Kullisaar T, Zilmer M, Mikelsaar M, Vihalemm T, Annuk H, Kairane C, et al. Two antioxidative lactobacilli strains as promising probiotics. International Journal of Food Microbiology. 2002;**72**:215-224

[21] Mudroňová D, Gancarčíková S, Nemcová R. Influence of zinc sulphate on the probiotic properties of *Lactobacillus plantarum* CCM 7102. Folia Veterinaria. 2019;**63**:45-54

[22] Shao L, Wu Z, Zhang H, Chen W, Ai L, Guo B. Partial characterization and immunostimulatory activity of exopolysaccharides from *Lactobacillus rhamnosus* KF5. Carbohydrate Polymers. 2014;**17**:51e56

[23] Ghosh T, Chattopadhyay K, Marschall M, Karmakar P, Mandal P, Ray B. Focus on antivirally active sulfated polysaccharides: From structure-activity analysis to clinical evaluation. Glycobiology. 2009;**19**:2-15

[24] Takeuchi A, Kamiryou Y, Yamada H, Eto M, Shibata K, Haruna K, et al. Oral administration of xanthan gum enhances antitumor activity through Toll-like receptor 4. International Immunopharmacology. 2009;**9**:1562-1567

[25] Patel S, Majumder A, Goyal A. Potentials of exopolysaccharides from lactic acid bacteria. Indian Journal of Microbiology. 2012;**52**:3e12

[26] Prete R, Alam MK, Perpetuini G, Perla C, Pittia P, Corsetti A. Lactic acid bacteria exopolysaccharides producers: A sustainable tool for functional foods. Foods. 2021;**10**:1653

[27] Senoner T, Dichtl W. Oxidative stress in cardiovascular diseases: Still a therapeutic target? Nutrients. 2019;**11**:2090

[28] Yang S, Lian G. ROS and diseases: Role in metabolism and energy supply. Molecular and Cellular Biochemistry. 2020;**467**:1-12

[29] Chiste RC, Freitas M, Mercadante AZ, Fernandes E. Superoxide anion radical: Generation and detection in cellular and non-cellular systems. Current Medicinal Chemistry. 2015;**22**:4234-4256

[30] Gulcin,̇ I. Antioxidants and antioxidant methods: An updated overview. Archives of Toxicology. 2020;**94**:651-715

[31] United Nations. Health and Population. Department of Economic and Social Affairs Sustainable Development; 2021

[32] Staudt A. Identification of Environmental Factors Critical to the Production of Exopolysaccharides by *Rhizobium tropici*. Indiana: University of Notre Dame; 2009

[33] Oleksy-Sobczak M, Klewicka E. Optimization of media composition to maximize the yield of exopolysaccharides production by *Lactobacillus rhamnosus*. Strains Probiotics and Antimicrobial Proteins. 2020;**12**:774-783

[34] Garcia-Garibay M, Marshall VME. Polymer production by *Lactobacillus delbrueckii* ssp. bulgaricus. Journal of Applied Bacteriology. 1997;**70**:325-328. DOI: 10.1111/j.1365-2672.1991.tb02943.x

[35] Fawole MO, Oso BA. Laboratory Manual of Microbiology. Owerri: Spectrum Books Limited Ibadan; 1998. pp. 71-81

[36] Lowry OH, Rosebrough NJ, Farr AL, Randall RJ. Protein measurement with the Folin phenol reagent. The Journal of Biological Chemistry. 1951;**193**:265-275

[37] Luang-In V, Saengha W, Deeseenthum S. Characterization and bioactivities of a novel exopolysaccharide produced from lactose by *Bacillus tequilensis* PS21 Isolated from Thai Milk Kefir. Microbiology and Biotechnology Letters. 2018;**46**(1):9-17

[38] Imran MY, Reehana N, Jayaraj KA, Ahamed AA, Dhanasekaran D, ThajuddinN,etal.Statisticaloptimization of exopolysaccharide production by *Lactobacillus plantarum* NTMI05 and NTMI20. International Journal of Biological Macromolecules. 2016;**93**:731-745. DOI: 10.1016/j.ijbiomac.2016.09.007

[39] Vijayalakshmi M, Ruckmani K. Ferric reducing anti-oxidant power assay in plant extract. Bangladesh Journal of Pharmacology. 2016;**11**(3):570-572. DOI: 10.3329/bjp.v11i3.27663

[40] Matsuzaki C, Takagaki C, Tomabechi Y, Forsberg LS, Heiss C, Azadi P, et al. Structural characterization of the immunostimulatory exopolysaccharide produced by *Leuconostoc mesenteroides* strain NTM048. Carbohydrate Research. 2017;**448**:95-102

[41] Ma JS, Liu H, Han CR, Zeng SJ, Xu XJ, Lu DJ, et al. Extraction, characterization and antioxidant activity of polysaccharide from *Pouteria campechiana* seed. Carbohydrate Polymers. 2020;**229**:115409. DOI: 10.1016/j.carbpol.2019.115409

[42] Re R, Pellegrini N, Proteggente A, Pannala A, Yang M, Rice-Evans C. Antioxidant activity applying an improved ABTS radical cation decolorization assay. Free Radical Biology & Medicine. 1999;**26**:1231-1237

[43] Seesuriyachan P, Kuntiya A, Hanmoungjai P, Techapun C. Exopolysaccharide production by *Lactobacillus confusus* TISTR 1498 using coconut water as an alternative carbon source: The effect of peptone, yeast extract and beef extract. Songklanakarin Journal of Science and Technology. 2011;**33**(4):379-387

[44] Salminen S, von Wright A. Lactic Acid Bacteria. New York: Marcel Dekker, Inc.; 1993

[45] Pham PL, Dupont I, Roy D, Lapointe G, Cerning J. Production of exopolysaccharide by *Lactobacillus rhamnosus* R and analysis of its enzymatic degradation during rolonged fermentation. Applied and Environmental Microbiology. 2000;**66**:2302-2310

[46] Brummer Y, Cui S. Food Carbohydrate: Chemistry, Physical

Properties, and Application. Boca Raton, FL: CRC Press Taylor & Francis Group; 2005. pp. 68-72

[47] Marie V, Collet F, Lanos C. Chemical and multi-physical characterization of agro-resources' by-product as a possible raw building material. Industrial Crops and Products. 2018;**120**:214-237. DOI: 10.1016/j.indcrop.2018.04.025

[48] Shanmugam S, Kumar TS, Selvam KV. Laboratory Handbook on Biochemistry. 2nd ed. Delhi: PHI Learning; 2019

[49] Pavia DL, Lampman GM, Kriz GS. Introduction to Spectroscopy. 4th ed. Washington: Brook/Cole Thomson Learning; 2009. pp. 13-101

[50] Grube M, Bekers M, Upite D, Kamisnka E. Infrared spectra of some fructans. Spectroscopy IOS Press. 2002;**16**:289-296

[51] Umeo SH, Faria MGI, Vilande SSS, Dragunski DC, Valle JS, Colauto NB, et al. Iron and zinc mycelial bioaccumulation in *Agaricus subrufescens* strains. Ciências Agrárias. 2019;**40**(6):2513-2522

[52] Arain M, Kamboh AA, Arshed MJ. Effects of selenium supplementation on hematological profile, gut microflora composition, in vitro biofilm formation assay and serum IgG concentration in goats. Pakistan Journal of Zoology. 2021:1-8. DOI: https://dx.doi.org/10.17582/journal.pjz/20200519140534

[53] Górska S, Maksymiuk A, Turło J. Selenium-containing polysaccharides—Structural diversity, biosynthesis, chemical modifications and biological activity. Applied Sciences. 2021;**11**:3717

[54] Benzie IFF, Strain JJ. The ferric reducing ability of plasma (FRAP) as a measurement of antioxidant power: The FRAP assay. Analytical Biochemistry. 1996;**239**:70-76

[55] Peter C, Wootton-Beard AM, Ryan L. Stability of the total antioxidant capacity and total polyphenol content of 23 commercially available vegetable juices before and after in vitro digestion measured by FRAP, DPPH, ABTS and Folin–Ciocalteu methods. Foodservice Research International. 2011;**44**(1):217-224